Homo Floresiensis: Diving Deep into our Evolutionary Past

Homo Floresiensis: Diving Deep into our Evolutionary Past

Lydia Sochan

Lilian Yeung

Romina Tabesh

Joylen Kingsley

Ariana Balassone

Sara Djeddi

Shahreen Rahman

Omar Hadi

Shannon Lin

Michaela Dowling

Ann Ping

Vedanshi Vala

with

Austin and Catherine Mardon

GM
PRESS

First Printing: 2021

Cover Design and typeset by Clare Dalton

ISBN 978-1-77369-674-4

E-book ISBN 978-1-77369-675-1

Golden Meteorite Press

103 11919 82 St NW

Edmonton, AB T5B 2W3

www.goldenmeteoritepress.com

Chapter 1: A Background on the Study of Fossils (Lilian Yeung)

A Brief History of Fossils' Excavation

The excavation of fossils has been of much interest to humans since the dawn of time. Fossils hold substantial information pertaining to our past as well as to the Earth and evolution. The study of fossils was not an intentional area of study, but rather was a result of accidental findings of fossil remains which led to interest and research into fossil remains. With the past lacking technological advances, much of fossil excavation was done through physical means using brushes, pickaxes, and shovels, consequently damaging remains through improper handling and accidental excavations in unsuitable conditions (Šejnoha et al., 2009). As more findings of fossilized remains began to occur leading to the interest in our ancient history, the fields of anthropology, archeology, and paleontology were born. With this expanding inquiry into the human past, researchers started searching for fossils and investigating them, making connections between our past and present, and leading to theories of evolution. As fossils have been tied to ancient lore and mysteries, much interest has been garnered in the research of fossils. Nowadays, fossil excavation is taught through many avenues, such as in museums, cultural or historic landmarks, and in the media. The media, through works such as the *Jurassic Park* movies and *Indiana Jones*, is especially powerful in drawing wonder and attention toward the fields of anthropology, paleontology and archaeology.

What Methods Are Used in the Study of Fossils?

The study of fossils is a complex and oftentimes incomplete journey as the process of fossilization includes the formation of irregularities and deformities. Remains of bones or organic materials will typically never be perfectly preserved, as a perfect fossilization process depends on a multitude of environmental factors such as the right pressure, temperature, humidity, and organic material. Hence, many fossils would include irregularities in their composition and also in the preservation process. In addition, fossils may also be deformed due to decomposition or other biological processes if sections of the remains were not preserved perfectly. The very nature of fossils entails many millions of years of diagenetic processes in addition to taphonomic processes (Lautenschlager, 2016). Diagenetic process or diagenesis is the process in which sediments are changed over time by physical and chemical means (Bjørlykke, 2013). These changes may be caused by water interacting with the sediment, in which sediments may be smoothed for example, or can also be caused by compaction and microbial activity (Bjørlykke, 2013). Taphonomic process is the fossilization process and is an important branch in the field of paleontology (Bjørlykke, 2013). As a result of these processes, fossil specimens will often display fractures, discolouration, along with the results of disarticulation, remineralization, and distortion once they are discovered (Lautenschlager, 2016). Furthermore, the process of extracting and collecting fossils may sometimes cause additional damage (Šejnoha et al., 2009). As such, advances are always being made to the techniques of acquiring fossils and restoring them for analysis in order to preserve them as much as possible (Lautenschlager, 2016). When examining fossils, researchers collect information on their size, colour, location, and composition to investigate their taxonomic positions and phylogenetic relationships (Lautenschlager, 2016). The fossils' appearance may tell great details about the possible ecology and behaviour of extinct organisms, whether they may be of human ancestry or from prehistoric animals, a conclusion which is often deduced wholly based on the preserved remains' morphology (Lautenschlager, 2016). Tools that are used to study fossils include both physical and digital methods such as computer-aided scanning, digital visualization, digitization techniques, and computational analyses (Lautenschlager, 2016).

An effective and straightforward way of digitally analysing specimens is the X-ray computed tomography, which is a high spatial resolution method that shows the anatomy of organisms in three dimensions for both *ex vivo* and *in vivo* applications (Gutiérrez et al., 2018). The imaging type tomography consists of using slices or sections that pass through a solid object, such as an organism or specimen, to provide imaging of the full structure of the specimen (Gutiérrez et al., 2018). A variety of methodologies may be used to achieve the slicing or sectioning of the organism. This method is widely used because of the benefit of imaging the full structure of the specimen without the requirement of invasive measures (Gutiérrez et al., 2018). With the use of this time efficient method, there is no destruction of the sample nor is there any deformation. Studies requiring continuous evaluations over time may find that tomography imaging is especially useful for such an endeavor. Moreover, X-ray computed tomography will have imaging of multiple sections and levels of the same sample providing images of different views (Gutiérrez et al., 2018).

A similar method to X-ray computed tomography is X-ray computed [micro-] tomography (microCT), which is commonly used to create three-dimensional models of the specimen (Immel et al., 2016). This method allows the exploration of the internal morphological structures, which allows one to distinguish between similar fossil remains. As microCT provides a virtual replica of fossils, researchers are able to share their findings with other researchers through a linked database (Immel et al., 2016). Through this database, it allows institutes and researchers to compile their findings, giving scientists the ability to collaborate with other institutions which is important to allow for new perspectives to be gleaned on research. In addition, the sharing of information is vital in tracking prehistoric remains and the regions where they once lived (Immel et al., 2016). Although microCT has been quite useful in the past, new research has shed light on the potential consequences that it may have on fossilized remains (Immel et al., 2016). Like any type of X-ray imaging, radiation is a common occurrence during the process (Immel et al., 2016). In living organisms, DNA strand breaks during X-ray imaging is not a cause for concern, as living organisms have repair mechanisms in place to fix damaged strands (Immel et al., 2016). Conversely, the same cannot be said for fossilized or mummified remains since its DNA strands have already been deteriorating throughout its entire fossilization process, and any additional damage to its DNA strands may impact the structural integrity of the remains

(Immel et al., 2016). As the method of microCT uses synchrotron sources, which is a form of electromagnetic radiation, in its retrieval of DNA, concerns have arisen regarding the possible deleterious effects of X-rays (Immel et al., 2016). Based on past scans examining dental structures where samples were exposed to synchrotron scans of high resolution, transitory darkening was observed on white or translucent enamel (Immel et al., 2016). This finding provoked concern over whether such effects were permanent or would lead to long-lasting damage (Immel et al., 2016). Although the transitory darkening may be removed easily with the use of little power and low energy UV rays, it is still a great cause for concern as the darkening shows that X-rays have a profound impact on specimens (Immel et al., 2016). Regardless, any method or tool that impacts and changes fossilized remains, albeit temporary, is still a concern for researchers as fossils cannot be reverted or recovered once changes have been made. Consequently, the use of X-ray imaging for specimens should be limited and used only if necessary, as the effects of X-rays are intense (Immel et al., 2016). Such effects include mainly electronic excitation along with ionisation that proves dangerous to specimens (Immel et al., 2016). These effects (i.e., DNA strand breaks, mutations, chemical modifications directed towards its bases, and structural changes) accumulate over time due to numerous X-ray scans along with the fossils' own natural taphonomic and deterioration processes (Immel et al., 2016). Accumulations of all these modifications, both natural processes and research processes, will accelerate the breakdown of fossils (Immel et al., 2016). Even though there is much benefit to using microCT, especially for dental structures and microstructures of fossils, the disadvantages are a cause for concern and continue to spark controversy among scientists (Immel et al., 2016).

CT scan is a safe and noninvasive measure that has its uses in not just the medical field, but also in the archaeological and anthropological field among others (Gutiérrez et al., 2018). Routine CT scans are typically advised for fossilized and mummified remains before any invasive measure may take place on the sample as this helps to preserve crucial internal information relating to its morphological data (Immel et al., 2016). There are many variations of CT scans which differ in qualities such as scanning time, resolution, the maximum and minimum specimen size it is able to scan, in addition to fiscal means (Gutiérrez et al., 2018). MicroCT scanners are primarily used to provide close up views of miniscule specimens ranging from about 3-50 millimetres by providing a high resolution image typically about 1–100 μm (Lautenschlager, 2016). The tradeoff between microCT scanners and medical CT scanners is that the microCT scanners can provide higher resolution images at the expense of limited specimen size while medical CT scanners can fit a larger range of specimen size with lower resolution imaging.

Besides tomographic methods (i.e., CT scans), there are also techniques that are surface-based scanning, which can provide adequate imaging for research studies that exclusively look at the external morphology (Gutiérrez et al., 2018). A negative aspect to surface-based scanning is that it cannot identify internal structures in imaging scans (Gutiérrez et al., 2018). An example of such a technique is laser scanning, a frequently used procedure where the specimen's external surface is probed using laser transmission (Lautenschlager, 2016). Laser scanning also varies depending on the system or model being used with some being able to attain sub-millimetre resolution in addition to being capable of scanning large specimens (Lautenschlager, 2016). The great advantage that laser scanning has compared to CT scanners is its mobility and efficient scanning times (Gutiérrez et al., 2018).

Another useful tool is photogrammetry, which is a safe, non-destructive, and non-invasive technique to provide digitalization of the fossil remains (Lautenschlager, 2016). This procedure is easy to use along with being cost-effective, making it a widely used procedure within anthropology (Lautenschlager, 2016). The resolution and quality of images depends on the make and model of the camera being used. Another technology within photogrammetry are unmanned aerial systems (UAS), such as drones (Waagen, 2018). Drones are a relatively recent technology used to capture topographic imaging to provide initial information on excavation sites (Waagen, 2018). With initial surveys of the land conducted through drones, researchers and engineers may have a better understanding of their excavation sites to devise a safe and proper excavation procedure to prevent damage to fossils, researchers, and their environment.

How Does the Field of Anthropology Connect to Other Scientific Disciplines?

The field of anthropology encompasses human cultures and societies in the past and their development (Wagner et al., 2016). It studies the beginnings of human civilization, human traditions, cultures, tools, behaviour, and society within these civilizations, and seeks to follow the development of such civilizations, as well as how they connect to today's societies and cultures (Wagner et al., 2016). When examining the broader field of anthropology, one can see that it closely ties into the field of sociology, the study of human societies, but specifically with regards to societal structure, functioning, and development (Wallerstein, 2002). While anthropology concentrates more so on the development of cultures and societies of past human civilizations, sociology tends to focus on recent, present societies and their structures (Wallerstein, 2002). Anthropology seeks to identify what makes people human, and thus, consists of four main branches which are the linguistic, archeology, biological, and sociocultural branches (Wallerstein, 2002). These are the four cornerstones that connect humans together creating society. Not only is anthropology an element of the social sciences and humanities, but it also connects to the physical sciences, such as geology, or earth science, and biology—especially where genetics is concerned. Genetics plays a massive role in shaping humans today, and fossils may hold DNA that can be compared to today's current human DNA, or even with DNA from other civilizations. The connection between biology and anthropology helps investigators understand what evolutionary changes may have occurred to result in our divergence from past human species (Gutiérrez et al., 2018). These evolutionary changes may include how and when different eye colours first appeared, the shape of our spine and bipedalism, along with skull shape and head sizes.

Anthropology also connects to history as well as economics and political science. Given that anthropology explores the development of human societies and cultures, it also incorporates economics and government (Wallerstein, 2002). Overall, anthropology has many ties to scientific fields, whether they be physical or social sciences. The field of anthropology cannot be looked at on its own without incorporating other fields of science to get a full comprehension of human societies, and how they once functioned and evolved. Anything that involves humans and their societies, as well as their cultures, can be connected to the study of anthropology—this includes topics such as fashion and its history.

However, the methods of excavation and fossil recovery used in anthropology closely align with methods in the field of paleontology (study of fossilized animals and plants) as well. Archeology, a direct branch of sociocultural anthropology, is responsible for finding and examining human remains. Archeology is a field that uses the same tools as paleontology, including digital imaging processes and preservation. Both fields have an overlap in the equipment and methods used to examine fossilized and mummified remains.

Conclusion

In conclusion, the study of fossils and its tools varies depending on the images and results that researchers are looking for. With imaging techniques such as microCT scanning, X-rays, laser scanning, and photogrammetry, they all work together to aid researchers in understanding their specimens. Although each method has its disadvantages, there are also advantages to using certain techniques to develop certain images. Imaging techniques are versatile and have many applications not just within the field of anthropology, but also within the fields of archaeology and paleontology. The use of these methods not only works to provide details for researchers, but also works to preserve fossils as much as possible, which is the goal of all these tools that are used within the anthropology field.

Chapter 2: The Discovery of *Homo floresiensis* (Romina Tabesh)

Human evolution is defined as the process by which modern humans originated from apelike ancestors. Many physical and behavioural traits shared by modern humans originate from these ancestors and have evolved over a long period of approximately six million years. Paleoanthropology is the study of human fossils in an effort to understand the early development of anatomically modern humans, a process known as hominization. Scientists have discovered and studied human fossils for many years. This field of study interests individuals today as it allows a better understanding of human origins and enables researchers to explore several aspects of human nature. This includes the ability of humans to acquire and understand language, cognitive skills and development, and different abilities of the human body.

Evolution: A Brief Summary

It is important to note that evolution occurs within a population and not within an individual. For example, a giraffe's neck would never 'evolve' to be longer, however, a population of giraffes could evolve to consist of individuals with longer necks over time in comparison to previous populations. This is because those with longer necks are able to reach higher trees and obtain food, whereas those with short necks simply do not survive. Over time, the population of giraffes that survive, which happen to be those with longer necks, will be naturally selected and pass on their "long neck genes' to their offspring. Similarly, mutations are occurring almost all the time within the human genome. These mutations are described as changes in the genetic material— which is called the DNA. Moreover, evolution occurs when there is accumulation of changes to a population's DNA. Furthermore, genes within the DNA have the genetic code— information necessary for a phenotype to be present. Genes affect the unique way in which the human body functions and behaviour develops, as well as many other factors like one's eye colour or height. This is the reason why characteristics which are genetically inherited can affect a population's survival and reproduction in a certain environment. Individuals with inherited genes that allow them to survive are deemed 'fit' and will be able to reproduce as a result of their fitness. Many changes made to the human genome enabled the species to survive, reproduce, and diverge over time.

The Discovery of *Homo floresiensis*

Discovered on the Island of Flores in Indonesia are one of the most recently uncovered human species called *Homo floresiensis*, nicknamed 'hobbit humans'. Their fossils were found on the Island of Flores in 2003 and from approximately 10,000 to 60,000 years ago (Karen L. Baab, 2012). Scientists also discovered stone tools made and used by *H. floresiensis* from 50,000 to 190,000 years ago (Karen L. Baab, 2012). Alongside Neanderthals, Denisovans, and *Homo sapiens*, *H. floresiensis* are one of the latest-surviving humans. The nickname 'hobbit humans' comes from several characteristics of *H. floresiensis* including their small stature. *H. floresiensis* stood about three feet six inches tall, had very small brains, and had teeth that appeared too large with proportion to their size. This species of human had no chin, a receding forehead, shrugged-forward shoulders, small legs, and large feet.

Though brain size usually correlates directly with cognitive abilities (e.g. thinking and language), *H. floresiensis* created many stone tools and used them to hunt rodents and elephants, as well as fight off predators. Surprisingly, this shows relatively strong cognitive abilities despite their brain size (Karen L. Baab, 2012). Within the Island of Flores where *H. floresiensis* resided, remains have been found in the cave of Liang Bua—this translates to 'cool cave'. Flores lies towards the eastern end of the Indonesian island chain and is separated from mainland Asia. It is believed that other animals reached this island by swimming or floating on debris, but paleontologists have yet to discover when or how *H. floresiensis* reached the island, as the nearest island is about six miles away.

While searching for evidence of early migration of *Homo sapiens* to Australia, an Australian-Indonesian team, Peter Brown, Michael Morwood and colleagues stumbled upon remains of a small-brained hominid now known as *H. floresiensis*. The remains discovered include a full skeleton with a nearly complete skull, a partial pelvis, a number of limb bones, and bones belonging to the hands and feet (LB1) as well as other parts belonging to at least eleven other individuals (Karen L. Baab, 2012). The LB1 specimen is believed to have stood about 1.06 m tall, which is comparable to a 3-4 year old modern human. LB1 is the only complete skull to have been recovered to date, but a mandible and several skeletal parts from a second individual (LB6) have also been found. Fossils of more than four other individuals were recovered, meaning this was a population of hobbits, and LB1 was not an anomaly (Karen L. Baab, 2012). Moreover, these remains were found to be approximately 60,000 to 100,000 years old and came from different levels. After the initial discovery of their remains, bones and teeth were found representing as many as twelve distinct *H. floresiensis* individuals in the cave of Liang Bua (Debbie Argue et al., 2017). So far, this is the only site where this species has been found. Stone tools were also discovered in several levels. Due to the fact that these remains are unfossilised and relatively young, researchers were looking to find mitochondrial DNA in order to trace ancestry. This is due to the fact that mitochondrial DNA is inherited solely from the mother, and hence stays the same over generations. So far, these efforts have been unsuccessful, but research still continues.

It is believed that early humans arrived on the Island of Flores at least one million years ago, but their method of arrival is unknown since the nearest island is approximately six miles away, across treacherous areas. Paleoanthropologists discovered several stone tools made and used by *H. floresiensis*. These tools are very similar to the ones found earlier on the island and throughout human evolutionary history. This includes Lower Paleolithic tools in Asia and Oldowan tools in Africa (Karen L. Baab, 2015).. Researchers also found evidence that these hobbit humans hunted an extinct type of elephant called Stegodon. Hundreds of Stegodon bone fragments were found within *H. floresiensis* occupation layers, some of which show butchery marks.

The Place of *Homo floresiensis* in the Evolutionary Tree

The position of *H. floresiensis* remains unclear on the human evolutionary tree and many controversies have arisen surrounding these specimens since their discovery. The era to which these fossils date has also been an interesting topic of study. The fossils range in age from 10,000 to 60,000 years, which falls within the range of modern humans elsewhere in the Old and possibly New World (Karen L. Baab, 2015). It is interesting to note, however, that only the younger layers above *H. floresiensis* fossils demonstrate modern human occupation at the Liang Bua cave.

Analysis of *H. floresiensis* fossils show ancestral traits that are preserved from ancient species, as well as derived traits which connect them to more recent ones. The skull represents a trait belonging to extinct species of the modern human genus *Homo*. The skeleton, on the other hand, is considered more primitive and in some ways aligns with fossils belonging to *Australopithecus afarensis* (Karen L. Baab, 2015). These are older, more primitive species. Altogether, such patterns are puzzling as they indicate the existence of differing origins—a population of small-brained species that existed 60,000 to 100,000 years ago, having a skull that closely resembles that of the much older *Homo habilis* or *Homo erectus*, and a skeleton with similar features to that of australopith species from at least 3,000,000 years ago.

The stone tools found in the Liang Bua cave, which are believed to have been made by *H. floresiensis*, are simple Oldowan-like tools, the most primitive types known in the archaeological record (Aida Gómez-Robles, 2016). These tools are similar in structure to others found at distinct sites on the Island of Flores which are nearly a million years old. In addition to butchering marks found on Stegodon bones, there are also burnt bones and pebbles, indicating intentional or perhaps accidental presence of fire.

When studying fossils, their morphology and age are taken into account in order to draw conclusions about the evolution of that species. In the case of *H. floresiensis*, the primitive morphology yet surprisingly young geological age of the fossils make it difficult to determine the exact evolutionary path of the species (Karen L. Baab, 2015). When first discovered, *H. floresiensis* were believed to be descendents of *Homo erectus*. Analysing the skeletal remains in more detail, however, later uncovered that the species has traits which more closely resemble archaic, and not Asian *Homo erectus* features. *H. floresiensis* were found to be more similar to *Homo habilis* and hominins than *Homo erectus*. This would suggest *H. floresiensis* are not actually descendents of *H. erectus*, but rather that the two species share a common ancestor.

Researchers have proposed several scenarios to explain the existence of *H. floresiensis* on Flores. One scenario links *H. floresiensis* with *Homo erectus*, who occupied Southeast Asia from about 1.5 millions years to perhaps more recently about half a million to 50 thousand years ago. Evidence to support this link are the shape of the brain, the low neurocranium, a flat and sloping forehead, thick cranial bones, a short and flat face, as well as other details of the skull anatomy of LB1 including an occipital torus and a mastoid fissure. The small body and brain size of *H. floresiensis*, however, do not fall within the expected range for that species. As a consequence of this discrepancy, it was proposed that these species be classified as a new species, *Homo floresiensis*, which are dwarfed descendents of *Homo erectus*. This classification is based primarily on the cranial evidence. Additional anatomical features of *H. floresiensis* support this, as they resemble that known of *Homo erectus*. This includes brain structure and shape, as well as the shape of the shoulder. In this scenario, it is believed that ancestors of *Homo floresiensis* made the treacherous water crossing and reached the Island of Flores. Over time, body size reduced. This is known as island dwarfing and has been observed in other large-bodied mammals such as primates, mammoths, and deer (Karen L. Baab, 2012). Island dwarfing is linked to reduced resource availability in the environment. It would be more advantageous in this case to be small-bodied, therefore needing less resources to survive.

In contrast to this, a different scenario is possible, where the population of *H. floresiensis* on Flores were the offshoot of a more primitive species known as pre-*erectus* hominin species which had small bodies and brains. This hypothesis is supported by evidence from the mandible—lower jaw— as well as the rest of the skeleton. The morphology of the teeth and mandible resemble those of *Australopithecus* and the earliest *Homo* species more closely than *Homo erectus*. In fact, very short legs (in comparison to the arms and feet) are common across apes and australopiths, but not *Homo erectus*. Overall, the morphology of *H. floresiensis* fossils appear to place the ancestry of this species earlier than *Homo erectus*.

To summarize the two scenarios presented, *H. erectus* ancestry would imply evolutionary convergence of postcranial structure with australopiths and early *Homo*. Reduction in body and brain size would suggest *H. floresiensis* are descendents of a *Homo erectus* ancestor. *Homo habilis* ancestry, on the other hand, is another possible scenario which is supported by several cranial features as well as dentognathic features. This type of evolution is known as parallel evolution and in this case implies ancestry between *H. floresiensis* and *Homo habilis*.

The absence of fossils of *Homo erectus* or any such species from island or mainland Southeast Asia is a major challenge to the theory that *Homo floresiensis* has a deeper ancestry than *Homo erectus.* Hominin fossil records prior to *Homo erectus* are found only in Africa. In addition to this, another complication is the poor fossil record of postcranial anatomy and morphology that is not thoroughly documented for pre-*erectus* species. Due to these challenges, it is unclear whether *Homo habilis* or such pre-*erectus* species are an appropriate model for representing the ancestor of *Homo floresiensis.*

Conclusion

Regardless of the origin of the population, it is evident that *H. floresiensis* faced isolation for a long period of time and underwent a phenomenon known as island dwarfing in the process. The result of this altogether is the 'hobbit' species that are now called *Homo floresiensis.* Studying human fossils allows researchers to study changes that took place in the evolutionary path of what is now known as the modern human. Moreover, the study of *H. floresiensis* fossils has enabled scientists to study changes in brain size, diet, and overall lifestyle of early human species. Placing these changes on a relative timeline allows individuals to expand their knowledge of human evolution and learn more about their roots as humans. Understanding evolution also helps scientists solve biological problems and several other issues impacting human lives today. Aside from studying fossils, researchers also study the evolutionary history of disease-causing genes. This allows the control of hereditary diseases in people. Overall, knowledge of evolution has helped and will continue to help improve the quality of human life. With continuous efforts and advancements being made regularly, new technologies can be created to enhance our way of life.

Chapter 3: Ethical Concerns in Human Fossil Research (Joylen Kingsley)

Ethics in Human Fossil Research: An Introduction

As humans we search our past to understand the direction of our future, such is the importance of archeology. In the 200 000 to 300 000 years since the evolution of *Homo sapiens*, people have woven a diverse web of religions, races, and beliefs, all shaped by the landscapes and natural phenomena native to the various corners of the world (Galway-Witham & Stringer, 2018). While understanding the intricacies of our diverse history gives us insight into the collective human past, that same diversity affects the ethical implications behind human remains research in modern society.

All professions operate on some form of ethical code that outlines the standard of behavior, ethics codes are objectively understood to be good, fair, and respectful towards the members they protect (Joannes-Boyau et al., 2020). The difficulty with ethics in human fossil research is the difference in belief systems around the world; distinct cultures, and the fact that societies approach death with opposing ideals, make it difficult to create a standard of ethics in human fossil research (Joannes-Boyau et al., 2020). There are many points in the duration of the archaeological process that are subject to debate in terms of ethics. What is the earliest point at which a body can be retrieved from its grave? Is it appropriate to uncover marked graves for the purpose of research? Who claims authority over archeological discoveries? These are all valid questions that must be explored. It is not enough to discover human history, researchers must also ensure that respect is delivered where due and remember that the remains belonged to humans who were once living and breathing, having their own experiences, cultures, and lives.

Recovering Human Remains: An Introduction

Ethics and archeology began to cross paths in the mid 1980s, when socio political issues became involved in archeological expeditions (González-Ruibal, 2018). Ethical dilemmas in archeology can be classified into three overlapping categories: practical, philosophical, and political (González-Ruibal, 2018). The question on recovering human remains encompasses all three categories, as do most ethical concerns in human fossil research. When is it appropriate to exhume a body, how much dignity must be given to the subject, and who claims authority over these discoveries (González-Ruibal, 2018)?

Recovering Human Remains: A Philosophical Perspective

From a philosophical perspective, Emmanuel Kant, the German philosopher, classifies autonomous beings as having dignity so long as they are willful and able to pursue their desires;

autonomous beings are worthy of respect, but the dead are not willful, nor do they have an innate goal to fulfill (de Tienda Palop & Currás, 2019). Although not explicitly stated, Scarre has expanded on Kant's philosophy to explain the innate wrongness of using human remains for anything other than what can be considered respectful (de Tienda Palop & Currás, 2019; Scarre, 2003). Despite the inanimate nature of human remains, they still hold traces of the living beings they once were. Although currently deceased, all human remains were once living beings, and the understanding is that a human who had desires and experiences in life would continue to have those same desires if not faced with death (de Tienda Palop & Currás, 2019; Scarre, 2003). For this reason, the ultimate clause of respect must be present when considering any archaeological expedition involving human remains.

Recovering Human Remains: A Practical Perspective

Practically, researchers must uphold responsibility to 'the people' regarding ethics in human fossil research. 'The people' not only encompasses the dead, but many groups who have concerns regarding the remains themselves, including but not limited to the archaeologists, indigenous communities, genetic investors, local governments, and the general public to name a few (Scarre, 2014). Every group has some sort of interest and place among an excavation site and claim to the discoveries, but the subsequent steps of the archaeological process is usually where interests vary, from analysis for research to reburial for religious purposes (Scarre, 2014). This begs the question as to which path is ethically correct. In most cases it is impossible to lay down a standard authority that encompasses all human fossil cases; interests vary from site to site and based on the situation calling for the exhumation (Scarre, 2014). Again, the common theme becomes dignity and respect, while interests may differ, respect offers a moral ground where all parties can agree.

Recovering Human Remains: A Political Perspective

Politically, the decision regarding recovery of human remains in the modern day must heavily involve the indigenous peoples affected by colonization. Colonization ethics have been implemented in mainstream archaeology and are extremely relevant in determining authority over remains (González-Ruibal, 2018). It is important for researchers to embrace the idea of 'the other.' There is a clear divide between researchers and the people native to archaeological sites (González-Ruibal, 2018). The ideal solution to political divides in ownership of human remains between these groups, is collaboration (González-Ruibal, 2018). While collaboration is easily implemented in theory, there are still many barriers in practice. The Native Graves Protection and Repatriation which passed in 1990 in the United States exhibits collaboration, where researchers are required to consult with local indigenous groups on the removal of remains and personal items, and where the local people can request the excavated subjects to be returned at any time (Scarre, 2014). This compromise is ideal for both parties, but also introduces another perspective. According to orthodox practices, researchers threaten scientific protocol by allowing open access to dig sites (Joannes-Boyau et al., 2020). The orthodoc perspective states that subsequent analysis of human remains is more important than the discovery of the remains themselves (Joannes-Boyau et al., 2020). Therefore, allowing public access may cause contamination or even destruction.

Recovering Human Remains: Conclusions

Ultimately, the ethical implications in most instances are situational. It is appropriate to exhume a body and claim authority over these discoveries, when it is respectful to do so. The interest of the people involved must be considered and collaborative efforts must be implemented. Researchers must keep in mind truth and justice when taking a scientific approach, and academic integrity and propriety when taking a humanitarian approach.

Displaying Human Remains: An Introduction

A community's relationship with and perspective on death heavily impacts the method of storage and display of human remains around the world (González-Ruibal, 2018). The concept of displaying human bodies has been a source of morbid curiosity for many generations. The approach to displaying remains varies depending on the age of the remains, the religious and political beliefs of the country, and the heritage of the country itself (González-Ruibal, 2018). The ethical implications also depend on the reason for displaying the remains; while the stereotypical display of bones is for anthropological and archeological reasons, there are also medical benefits in human fossil research; all of this must be considered in the overarching branch of ethics (Sallam, 2019). As indicated earlier in this chapter, ethical dilemmas in archeology can be classified into three overlapping categories:

practical, philosophical, and political. The ethical debate regarding the exhibition of remains primarily encompasses a philosophical and political perspective.

Displaying Human Remains: A Philosophical Perspective

The attitude towards death relies on the heritage and the religious beliefs of the region where the remains were discovered, the beliefs of the researchers, the beliefs of the indigenous peoples, and the beliefs of those who choose to display the fossils. Western cultures, primarily North Americans, tend not to associate themselves with the topic of death, but rather view death from a clinical perspective (Overholtzer & Argueta, 2018; Swain, 2002). In contrast to this, Eastern cultures, including many Asian countries, are more likely to engage remains with a more personal connection (de Tienda Palop & Currás, 2019). Ironically, there are many schools of thought even within the archeological community (de Tienda Palop & Currás, 2019). While Southern European archeologists are more neutral regarding their stance on death, emphasizing the biological nature of their discoveries, Central European researchers are more likely to humanize their discoveries, prioritizing their moral treatment (de Tienda Palop & Currás, 2019). The orthodox perspective on presenting remains has been that of the English, who tend to distance themselves from the dead (Overholtzer & Argueta, 2018; Swain, 2002). They generally believe that there is no connection between the skeleton and the person who they once were in life (Overholtzer & Argueta, 2018; Swain, 2002). This attitude allows displaying remains to be more acceptable (Overholtzer & Argueta, 2018; Swain, 2002).

Indigenous groups to the United Kingdom are either indifferent or even favor the display of human remains, although it must be noted that colonization was not a devastating issue from a historic British perspective (Overholtzer & Argueta, 2018).

Displaying Human Remains: A Political Perspective

While Britain remains firm in their beliefs, much of the world has started to transition towards a less visual form of archeological research (Swain, 2002). Since its introduction in 1990, the United States Native American Graves Protection and Repatriation Act has greatly influenced the Western perspective on displaying remains (Overholtzer & Argueta, 2018). Much of Europe, the Americas and Oceania have started to turn away from displaying remains, and while the topic has not been widely explored by the Asian and African continents, the discussion always redirects to the importance of decolonizing the approach to burial and displaying remains (Overholtzer & Argueta, 2018). While the argument of Western societies primarily focuses on demonetizing cultural heritage and decolonization, giving the affected community the decision to display remains, the opposing perspective must also be considered (González-Ruibal, 2018). When properly sourced and with collaboration among indigenous communities, the remains may be used to present culture to the world, and contribute to younger generations in the community, not to mention the scientific implications of displaying remains in published papers (González-Ruibal, 2018). For example, the Xaltocan community in central Mexico wishes to utilize the remains of their ancestors to bring justice to their own political status and to increase the presence and understanding of their social identity (Overholtzer & Argueta, 2018). While there were challenges from a political and individual level, the intention to display the human fossils remains the same, based on the opinions of the community.

The medical community also introduces a more complicated aspect to the debate of displaying human remains. The purpose of medical display can be classified into pathological and clinical perspectives. The clinical perspective tends to take a more holistic approach with regard to patients, examining the history and family connection beyond the medical scope (Sallam, 2019). Opposingly, the pathological perspective focuses on the lesion or wound itself, almost erasing the identity of the fossils (Sallam, 2019). Presentation for medical purposes focuses on lighting and frame, to bring attention to the pathology itself (Sallam, 2019). While significant to medical understanding, the method of display redirects attention away from the individual on display, and as a byproduct the sanctity of their identity (Sallam, 2019). The solution proposed by the International Council of Museum's code of ethics states that human remains must be displayed in a respectful and professional manner; the remains should only be displayed if the same message cannot be expressed by an equally effective method without a display (Sallam, 2019).

Displaying Human Remains: Conclusions

Despite the obvious differences in scenarios between retrieving human remains and displaying them, the common theme is respect for human life and giving credit and authority where due. Reparations and reburial must also be considered when analyzing the ethics behind displaying human remains (Overholtzer &

Argueta, 2018). With respect to decolonizing fossil research, many native people believe that reburial is vital in the process of death (Joannes-Boyau et al., 2020). Academics on the other hand believe that reburial increases decay. It is important to consider the future of the human remains when displaying them (Overholtzer & Argueta, 2018). The debate between opposing factions regarding reparations and reburial is another topic that must be further explored, to fully acknowledge the rights of the parties involved in fossil research.

Ethics in Human Fossil Research: Conclusions

There are many stages to consider in human fossil research, from the discovery to display to reburial. While this chapter may not have answered the correct way to approach unearthing human remains or the procedure on presenting them, it does touch on the sanctity of death from various perspectives and the ever increasingly important decolonization of ethics in archaeology. As the world becomes more aware of cultural injustices, the same principles must be applied in archaeology. Ultimately, while there are no universal laws, there are moral obligations that all researchers should aspire to follow (The Norwegian National Research Ethics Committees, 2016). There should be equal respect for the dead with no bias of origin; when decisions are being finalized there should be consideration for the descendants and affected community where applicable (The Norwegian National Research Ethics Committees, 2016). All projects must be considered for their feasibility and all materials must be treated with care for both scientific and moral purposes; the project must be considered in the proper context of the region, with an in-depth understanding of the consequences (The Norwegian National Research Ethics Committees, 2016). And finally, while it is impossible to create laws for every specific situation, any reasonable existing laws must be taken into account (The Norwegian National Research Ethics Committees, 2016). It is important to remember that despite their significance to science, human remains were once living and loved, and therefore they have a certain degree of moral status. As researchers and fellow human beings, we must acknowledge the contributions of our ancestors, whose relics and remains provide us with a rich understanding of our history. At the very least for this reason, their remains must be respected ethically (de Tienda Palop & Currás, 2019).

Chapter 4: A Background on the Study of Human Evolution (Ariana Balassone)

Combining Genetic and Fossil Evidence

There is much to be discovered regarding the origin and evolution of modern humans. Genetic and fossil evidence have provided great advances in this field of research. Fossil evidence in the form of skeletons, teeth, stone tools, figurines, paintings, footprints, and other traces of human behaviour present insight about human evolution (Gasperskaja & Kučinskas, 2017).

Fossils allow researchers to study specific changes that occurred in human mobility, brain and body size, diet, and other prospects of prehistoric human life over millions of years (Gasperskaja & Kučinskas, 2017). On the other hand, genetic evidence enables researchers to study how closely related modern humans are to other primates, and can indicate how humans migrated in prehistoric times.

DNA is the genetic code carried in every single cell in the human body, and every person is 99.9% identical in their genetic makeup (Gasperskaja & Kučinskas, 2017). Interestingly, the other 0.1% accounts for differences in our phenotype and contain important information about evolution (Gasperskaja & Kučinskas, 2017). Researchers can study how human DNA differs geographically to gain an understanding of human evolution (Gasperskaja & Kučinskas, 2017). By combining genetic and fossil evidence, the mysteries of modern human origins and evolution can be solved.

Natural Selection

Why do people around the world look so different? Why are some people more susceptible to certain diseases than others? The answer lies in natural selection and genetic diversity. Recall, human DNA only differs by 0.1%, and these variations are what make humans unique, they differ geographically because of natural selection (Novembre & Di Rienzo, 2009). If an advantageous allele is the result of a unique mutation, it will originate in a single geographic location and spread outwards in a wave-like pattern (Novembre & Di Rienzo, 2009). The correlation between geographic locations and the spread of alleles arises when selective pressures differ in certain areas due to environmental factors (Novembre & Di Rienzo, 2009). Essentially, certain alleles persist in specific locations where they are favoured because of geographic factors. Whether the allele spreads or not depends entirely on how selective the advantage of the variant is, and the dispersal patterns of a population.

There are various methods used in the detection of these spatial identifications of selection. The investigation of the spatial patterns of selection can provide researchers with knowledge of how humans have evolved in response to selective pressures (Novembre & Di Rienzo, 2009). It is known that humans are extremely genetically similar, yet geographic patterns in DNA are shown in many heritable attributes (Novembre & Di Rienzo, 2009). These traits are conditions such as disease risk, pathogen resistance, drug response, and symptom severity (Novembre & Di Rienzo, 2009). People are phenotypically distinct depending on their geographic location. Skin pigmentation, body mass, hair colour and type are all due to selective pressures (Novembre & Di Rienzo, 2009). Which genes suggest the occurrence of natural selection? Which genes are associated with the current patterns of geographic genetic diversity? Answering these questions is important to the study of the genetics behind human evolution and phenotypic diversity.

The Clusters and Clines Pattern of Human Variation

There is evidence of at least five genetic clusters based in continents around the world (Novembre & Di Rienzo, 2009). The change in phenotype of the population in these continents tends to be gradual and they tend to have ancestral links to both sides of the separating boundaries (Novembre & Di Rienzo, 2009). There are several scientific explanations for these geographic patterns of phenotypic variation. For example, one model known as isolation by distance, suggests that gene flow between populations is dependent on how geographically close two populations are. (Novembre & Di Rienzo, 2009). This can easily be shown by comparing the phenotypic similarities between two geographically adjacent locations, such as Poland and Germany, as well as Spain and Portugal. Individuals from Poland and Germany commonly have lighter eyes and fairer skin tones, whereas Spanish and Portuguese people generally have darker eyes and complexions.

Secondary contact is another model that provides evidence of these geographic relationships associated with phenotypic traits. Two allopatrically isolated populations (populations that have been geographically isolated to an extent that prevents gene flow) reunite again promoting exchange of alleles through reproduction (Novembre & Di Rienzo, 2009). A population cline (a gradient of morphological change defined by geographical transition) is formed as a result of genetic drift, where an ancestral population emerges outwards into uncolonized locations and forms new sub-populations (Novembre & Di Rienzo, 2009). These new populations arise because of different environmental factors, resulting in different selective pressures. As a result, alleles that are at a low frequency in the original population can rise to rapid levels in the new population (Novembre & Di Rienzo, 2009). This creates what is commonly known as the bottleneck distribution of alleles (a steep decline in the size of a population due to unpredictable environmental events). Population clines and clusters are consistent with the observed patterns of evolution.

Detecting Selection Methods

A fixation index (F_{ST}) is a statistic used to measure the amount of genetic differentiation among populations (Novembre & Di Rienzo, 2009). A higher F_{ST} statistic indicates that there was an outwards expansion of a globally advantageous allele or there are local selective pressures. Essentially, a specific allele provides a selective advantage so it increases in frequency. This suggests that one population has undergone selection at a certain loci (a fixed region on a chromosome where a certain gene is) that another population has not, and so the two groups have a higher level of differentiation (Novembre & Di Rienzo, 2009). On the other hand, alleles maintained at a neutral frequency are expected to show less differentiation, and thus have a lower F_{ST}.

A popular and efficient approach to detecting spatial patterns of selection involves defining groups of SNPs (single nucleotide polymorphisms), and then determining whether the F_{ST} value is statistically significant for two different sets of these SNPs (Novembre & Di Rienzo, 2009). This approach is used to identify the individual geographic point of such adaptations, such that when selection is prevalent, the loci under selection will demonstrate strong degrees of differentiation relative to the neutral loci (Novembre & Di Rienzo, 2009). Thus, such SNPs can be identified using an outlier approach by comparing the proportion of SNPs with very high values of the F_{ST} statistic to the proportion of SNPs with neutral F_{ST} values. In fact, there are SNPs in the skin pigmentation genes that have been identified as advantageous alleles by selective pressures (Novembre & Di Rienzo, 2009). However, this method has some important limitations to take into account. Background selection occurs if natural selection repeatedly removes new deleterious mutations from a population of loci. This will increase the rate of differentiation and result in higher than normal F_{ST} values (Novembre & Di Rienzo, 2009). Thus, varying classes of SNPs in coding regions need to be compared to demonstrate a pattern of inheritance in certain populations.

Adaptive Variation

A new allele can be introduced into a population via mutation, dispersal from a neighbouring population, or through standing variation (when a segregating variant becomes favourable) (Novembre & Di Rienzo, 2009). The new allele either increases in frequency in a population because it is advantageous, or it is eliminated from the population because it serves no purpose or is unfavourable.

Examples of Natural Selection in Humans

Some of the best known examples of natural selection in humans are due to multiple underlying mutations that confer an advantage. This is seen in lactase persistence, skin pigmentation, and malaria resistance polymorphisms.

Lactase persistence is the ability to digest lactose, a sugar found in cow's milk. Humans did not originally have the ability to consume the milk of another mammal, however, in Europe the domestication of cattle grew quite popular (Hedrick, 2011). It has been hypothesized that the allele for digesting lactose originated in Europe (Hedrick, 2011).

The highest spread of malaria is within Africa because the warm climate provides suitable living conditions for the mosquitoes that act as transmission agents. Moreover, the high population density in Africa greatly promoted the spread of infectious diseases, and enhanced the use of pathogens as natural selection agents (Hedrick, 2011). Malaria is not a surprising selective force as it is one of the population's oldest diseases and it has caused numerous deaths (Hedrick, 2011). When a mutation arose that could combat the contraction of malaria it persisted within the population because it provided a selective advantage under certain environmental conditions.

Sickle cell anemia is a genetic condition in which the red blood cells are elongated and therefore they cannot pass through the capillaries or carry oxygen efficiently (Hedrick, 2011). The geographic distribution of the sickle-cell mutation associated with the beta hemoglobin gene is limited to Africa and certain regions within Asia where malaria is endemic. Individuals with the sickle cell mutation are less susceptible to malaria which is why the mutation persists in tropical regions where malaria is more common (Hedrick, 2011). Essentially, it is an X-linked recessive trait (a mode of inheritance in which the gene is carried on the x chromosome such that males as well as homozygous females will always express the associated phenotype) and carriers can protect themselves from contracting malaria because the parasite that causes the disease cannot infect sickled red blood cells (Hedrick, 2011). Heterozygotes of the sickle cell mutation do not experience the compromised blood flow that homozygous sickle-cell patients would endure, because only a portion of their cells are sickled, and they are also protected against contracting malaria. (Hedrick, 2011). This explains why this mutation is advantageous and persistent in regions where malaria is endemic.

Genetic Analysis Methods and Technologies

The human genome has been sequenced, however the challenge is to interpret the function of genes and gene products, which requires many genetic tools. Throughout the years there have been numerous technologies and tools used in genome analysis (Gasperskaja & Kučinskas, 2017). These technologies have improved over the past decade as there have been advances in high throughput methods. Advances include real time polymerase chain reactions, next generation sequencing, and mass spectrometry (Gasperskaja & Kučinskas, 2017). These scientific methods of study can help researchers answer questions in the fields of genomics, proteomics, epigenomics, and interactomics, which are essential to answer further questions in the dynamics of biological processes at both a cellular and structural level (Gasperskaja & Kučinskas, 2017). Each scientific method in genetic analyses has strengths and limitations that are important to consider, and depending on the area of study, some methods are better than others. Although it is very difficult to entirely understand the function of genes and gene products, knowing this information can promote revolutionary progress in the study of human evolution and natural selection. If researchers understand the function of human genes, observing how these genes vary across the world will provide insight on how prehistoric humans lived and evolved into modern humans.

Common Genetic Tools

It is important to discuss a few techniques and methods used in genetic analysis and to understand how these genetic tools work. As stated previously, the human genome is 99.9% identical, with a 0.1% difference that accounts for phenotypic variants in a population (Cavalli-Sforza & Feldman, 2003). These differences in the human genetic code can range from small SNPs to large chromosomal aberrations, and there are specific sequencing technologies used to detect these variants (Gasperskaja & Kučinskas, 2017). There are three types of genetic tests used to detect abnormalities in chromosomal structure, SNPs, protein function, and DNA sequence. Cytogenetics involves the observation of chromosomes to detect any abnormalities. For example, the most common technology used to detect large chromosomal aberrations (changes to the structure or number of chromosomes) is a karyotype analysis using the GTG banding technique, where condensed chromosomes are stained to produce a visible karyotype with a unique pattern of light and dark banding. The distinct banding on each chromosome is used to identify its structure. Researchers can easily determine the number of chromosomes and detect any type of chromosomal rearrangement using the GTG banding method. Molecular testing is used when the desired DNA sequence is known.

For smaller variations in the DNA sequence such as mutations, direct DNA testing is the most effective approach. Such molecular technologies include polymerase chain reaction (PCR), comparative genomic hybridization (CGH), and DNA hybridization (Gasperskaja & Kučinskas, 2017). During a PCR procedure, a specific segment of DNA is amplified via repeated cycles of denaturation, annealing, and elongation. A microarray-based comparative genomic hybridization is used to detect copy number variations in DNA (Gasperskaja & Kučinskas, 2017). First probes, which are strands of DNA, are prepared, and can vary in size. DNA is extracted from both a test sample and a reference sample and then denatured so that it is single stranded (Gasperskaja & Kučinskas, 2017). Both strands of DNA are fluorescently labelled with two separate dyes and the DNA strands are mixed together and applied to an array, where they attempt to hybridize with the single stranded probes. Then the proportion of fluorescently labelled patient DNA is compared to normal reference DNA (Gasperskaja & Kučinskas, 2017).

Moreover, DNA microarray analysis is used to detect the level of gene expression in a cell of interest. Molecules of mRNA hybridize to DNA probes and a computer detects the amount of mRNA bound to each array to determine the level of gene expression. Furthermore, biochemical tests analyze protein activity or protein quantity (Gasperskaja & Kučinskas, 2017). A popular technique is high performance liquid chromatography (HPCL), in which amino acids are passed through a column and treated with a fluorescent dye. A plot of absorbance as a function of time is computed to measure protein quantity. The type of genetic testing performed depends on the type of aberrancy in DNA being measured.

Fossil Evidence for Evolution

Aside from looking at the genetic evidence for evolution that lies within global phenotypic variation, there are also fossils that provide researchers with insight regarding prehistoric humans. In one study, researchers specifically examine fossils and the evolution of human brain and body size. First, what factors contribute to the size of a human, or any living thing for that matter? Energy requirements, social organization, relative brain size, locomotion, and various other morphological and life history characteristics are all involved in relative body size (Kappelman, 1996).

In a recent study, researchers examined the relationship between orbital area, which is the area surrounding human eye sockets, body mass, and the evolution of relative brain size. They then formulated equations to predict the body mass for the fossil hominid crania in which the endocranial volume is known (Kappelman, 1996). For fossil hominid crania (human skull), orbital area can be used to estimate body mass, and these body mass predictions can then be connected with previously measured endocranial volumes (a cast taken from inside the braincase) to determine approximate brain size (Kappelman, 1996). This approach is strong because it is dependent on a cranial indicator such as orbital area, which allows a direct calculation of proportionate brain size for fossil crania (Kappelman, 1996). Additionally, because most fossils are generally characterized by cranial characteristics, this method of comparison avoids the issue of sorting isolated fossils to a particular taxon (species) when predicting body mass. Furthermore, sorting male and female crania is often possible, so this approach also enables researchers to examine disparities in body mass or brain size that can be associated with sex (Kappelman, 1996). Scientists have found a strong correlation between body mass and orbital area, and from this they have concluded that modern humans have bigger brains and smaller bodies than prehistoric humans (Kappelman, 1996).

Selective Pressures

Human fossils have shown that modern humans have larger brains and hence, a more advanced emotional quotient (EQ), why would this be favoured? For one, larger brains correlate with intelligence, and one must be smart in order to ensure survival (Kappelman, 1996). Perhaps, a social selective pressure existed in which higher levels of cooperation between groups led to more efficient communication, and a higher EQ is required for interaction, hence the selection for individuals with bigger brains (Kappelman, 1996). In turn, a decreased body size could be observed because reduction in skeletal robusticity would have produced sufficient energy savings that promoted the maintenance of larger brains (Kappelman, 1996). Social selective pressures are most likely the reason for the selection of bigger brains because a high EQ is necessary for humans to communicate and hence reproduce (Kappelman, 1996). More intelligent humans with bigger brains were likely the only ones reproducing and passing on their alleles, until the population consisted of predominantly larger-brained humans.

Conclusion

Human evolution is a complex process that takes place over millions of years. One of the most important contributing factors of evolution is natural selection, the distinctive survival of individuals who bear a genetic advantage (Cavalli-Sforza & Feldman, 2003). Genetic diversity and human fossils are evidence of human evolution. Every person is phenotypically unique due to underlying genetic polymorphisms that are advantageous in certain scenarios. For instance, the sickle cell mutation persists solely in tropical regions where malaria is endemic because it provides protection against the Plasmodium parasite, which cannot infect sickled red blood cells (Hedrick, 2011). Furthermore, evolution is demonstrated within human fossils. Through the dating of discovered human fossils, researchers have concluded that humans have evolved to have bigger brains because intelligence is a necessity in effective communication and social interaction between kin. Certain alleles are favoured and will excel in a population where they provide a selective advantage (Cavalli-Sforza & Feldman, 2003). Recently there have been great advances in high-throughput methods of genetic analysis that have contributed transformative progress in the study of human evolution, however every technology has its disadvantages that need to be taken into account (Cavalli-Sforza & Feldman, 2003). Despite the limitations in the accuracy of the technologies used in genetic analysis, or in the prediction calculations used in fossil studies, the combination of human fossils and genetic testing have made revolutionizing advances in the study of evolution.

Chapter 5: The Speciation Debate: *Homo Floresiensis* (Sara Djeddi)

Introduction

In 2003, the discovery of a new possible human species, *Homo floresiensis*, sparked controversy with regard to their evolutionary status (Heteren, 2011). The fossils found in Flores, Indonesia belonged to hominids with a small stature, or real life 'Hobbits' (Baab, 2012). Throughout time, many theories have been proposed to explain the morphology and recency of the fossils. A number of researchers have stated that *H. floresiensis* are a valid new species, attributing their height to island dwarfism (Baab, 2012). Others have been more skeptical, claiming that the discovered bones are rather of a pathological modern *Homo sapiens* (Groves, 2007). This type of dispute is not uncommon; in fact it has happened following the discovery of many important human fossils (Groves, 2007). However, it is the recency of the *H. floresiensis'* disappearance that makes this valid-species debate interesting. The remains were found in the Liang Bua cave by Australian and Indonesian researchers, and hence named the LB1 skeleton (Falk et al., 2007). LB1 is presumed to have been an adult female that stood at about 3 feet tall (Falk et al., 2007). The bones belonging to LB1 date to as recently as 12 000 years. In contrast, *Homo erectus* fossils range from 2 million to half a million years old (Baab, 2012). This is important because the close proximity in time to modern *H. sapiens* could be supportive of the pathological hypothesis.

What is a Species?

Before beginning a discussion on whether or not *H. floresiensis* is a distinct species, it is important to understand what a species is, and why defining it is important. From a biological perspective, a species is a collection of organisms that are capable of interbreeding and producing fertile offspring (Aldhebiani, 2018). In terms of morphology, the focus is rather on the uniqueness of structural features, known as morphological characteristics (Aldhebiani, 2018). As of 2021, there are approximately 21 human species that have been discovered ("What does it mean to be human", 2021). However, it is difficult to simply give one number, as it differs depending on the source. All of this plays a key role in understanding the evolutionary path that led to modern *H. sapiens*. Human fossil discoveries and their respective place in the timeline are important in creating an accurate phylogenetic tree of the human species. Human evolution is not a concept that is completely understood; there are still many theories and debates surrounding it. This is why classifying new discoveries are crucial, as they can provide more information on human history.

Possible Link to *H. erectus*

When the fossils in the Liang Bua cave were first discovered, it was hypothesized that *H. floresiensis* was linked to another human species known as *H. erectus* (Groves, 2007). This assumption was primarily based on the similarities in the morphology of the remains (Groves, 2007). Both *H. erectus* and *H. floresiensis* have similar brain and shoulder shapes, flat and short faces, receding foreheads, as well as comparable cranial anatomy (Gordon et al., 2008). However, the major difference between the two is the smaller brain and body size of *H. floresiensis*, including shorter brow ridges and shortened lower limbs (Gordon et al., 2008). Essentially, it was thought that the *H. erectus* species, which was present in Southeast Asia almost 2 million years ago, evolved shorter heights by means of island dwarfing (Heteren, 2011).

In 2004, one year after the discovery of the remains, the theory of island dwarfism was introduced (Heteren, 2011). Although it has recently been shown to be unlikely, at the time it seemed like a plausible explanation (Heteren, 2011). The idea was that *H. erectus* were the ancestors of *H. floresiensis*, and they were exposed to long-term isolation on the island of Flores resulting in dwarfing to become *H. floresiensis* (Brown et. al., 2004). The reduction in body size is related to the limited resources in island environments, making it more advantageous to have a smaller stature (Baab, 2012). Many cases of island dwarfism have been reported among large mammals that have been isolated (Baab, 2012). One example of this is *Stegodon*, a now extinct elephant-like species that also exhibited this phenomenon (Heteren, 2011). Interestingly, *Stegodon* remains were found in the Liang Bua cave, close to the aforementioned *H. floresiensis* fossils (Heteren, 2011).

The possibility that there is a link between *H. erectus* and *H. floresiensis* was shown to be improbable in 2005 after comparison of their body size and brain size (Falk et al., 2005). Using CT scans, it was determined that the endocranial volume of LB1 is approximately 400 cm^3 compared to the 1200 cm^3 of modern human brains (Falk et al., 2005). The large discrepancy in endocranial volume is not something that can be explained by island dwarfism (Martin et al., 2006). This is because the size of the brain tends to decrease less than the size of the body (Martin et al., 2006). In the case of *H. floresiensis*, there exists a significant decrease in both, demonstrating that the *H. erectus* theory may not be plausible after all.

Pathological *Homo Sapiens* Theories

Over the years, it has been suggested that *H. floresiensis* is not a distinct species, but rather a diseased or pathological modern human. The speculation began with a group known as the Rampasasa pygmies (Groves, 2007). These individuals have a short stature, of about 150 cm, and live close to the Liang Bua cave where the *H. floresiensis* fossils were discovered (Groves, 2007). A few key characteristics of the pygmies are their receding chins as well as asymmetrical skulls, alluding to pathological abnormalities (Groves, 2007). It was thought that *H. floresiensis* were the ancestors of the Rampasasa people, but this was disproven in the following years. The pygmies' receding chin was compared to that of LB1, showing to be within modern human range (Jacob et al., 2006). However, a major difference lies within the structure of the *H. sapiens* symphysis, a joint connecting two bones, which *H. floresiensis* lacks completely (Groves, 2007). Additionally, neither modern humans nor Rampasasa pygmies have yet to exhibit the small Hobbit heights of about 100 cm (Groves, 2007). Rather, the very small stature of *H.*

floresiensis has been attributed to pathological conditions such as microcephaly or Laron syndrome (Groves, 2007).

Microcephaly, a disease causing underdeveloped brains, has been suggested as an explanation for the small neurocranium of LB1 (Groves, 2007). In 2006, it was concluded that the body size of LB1 was too small in relation to its brain size for it to be a dwarf species, and instead it was more likely suffering from microcephaly (Martin et al., 2006). Additional research was conducted comparing the skull of LB1 to both modern microcephalics as well as microcephalic skulls that are thousands of years old (Heteren, 2011). In both of these cases, the researchers at the time stated that it was very probable that LB1 suffered from microcephaly due to the resemblance of the craniums (Groves, 2007). This conclusion was based on the evidence suggesting similarities in the shape of the skull and size of the brain (Groves, 2007).

The discrepancy between LB1 and microcephalics is rather in the details of the brain and skull (Falk et al., 2007). In 2007, it was found that the *H. floresiensis* brain has developed frontal and temporal lobes, in contrast to microcephalic brains which do not (Falk et al., 2007). Although the endocranial volume of an individual with microcephaly is less than 700 cm^3, there are not many cases where this number reaches less than 400 cm^3 (Groves, 2007). Those with microcephaly are also somewhat small in stature; however, the similarities to *H. floresiensis* are not enough to confirm the correlation (Groves, 2007). Furthermore, the observations made by the earlier researchers were based on the dimensions of the brains rather than their complete cranial morphology (Heteren, 2011). After the initial suggestions, additional studies have been conducted that show that microcephalic brains do not actually resemble that of LB1 beyond the similar brain size (Heteren, 2011). Specifically, a more recent study in 2016 established that the skull of LB1 has no evidence of microcephaly whatsoever, thus furthering the narrative that they are in fact distinct species rather than microcephalic modern humans (Balzeau & Charlier, 2016).

The other pathological condition used to explain the small stature and brain size of LB1 is Laron Syndrome. Due to growth hormone insensitivity, those with Laron Syndrome have malfunctioning receptors that do not respond properly to the growth hormones produced in the body (Baab, 2012). This results in individuals with small skulls and short heights, similar to *H. floresiensis* (Baab, 2012). Laron Syndrome is also more common in populations that take part in interbreeding, which could have been the case on the island of Flores (Laron, 2004). A study conducted in 2007 suggested that *H. floresiensis* are *H. sapiens* with a mutated GH receptor, which causes Laron Syndrome (Hershkovitz et al., 2007). However, it was later determined that the study was missing some information, and that LB1 showed no evidence of having Laron Syndrome (Heteren, 2011).

In terms of their skulls for example, LB1 is thought to have a sloping forehead that slants back rather than forward (Baab, 2012). Their foreheads consequently make their faces appear large compared to their cranium (Baab, 2012). This is entirely different from Laron Syndrome patients who have a prominent forehead that protrudes outward (Baab, 2012). These patients also have small faces in relation to their heads due to underdeveloped bones (Baab, 2012). Another example is that LB1 has big feet whereas Laron Syndrome patients have shortened feet and hands (Hetern, 2011). There are also many features of LB1 that were not included in the 2007 study that deviate from the characteristics of Laron Syndrome (Baab, 2012). Moreover, similarly to microcephalic individuals, those with Laron syndrome do not have an endocranial volume as low as LB1's 400 cm^3 (Heteren, 2011). Therefore, once again, the morphological characters of LB1 do not fit the criteria to be considered a diseased modern human.

New Discoveries

The previous theories stating that *H. floresiensis* are pathological modern humans only took the remains of LB1 into consideration. Over the years, additional *H. floresiensis* fossils belonging to various individuals have been discovered (Groves, 2007). This new evidence is crucial in the determination of whether or not *H. floresiensis* are a distinct species of human. After the discovery of LB1, the remains of LB2 were discovered as well as "LB3, a tiny ulna, LB4, a child represented by a radius and tibia, LB5, a vertebra and a metacarpal from an adult, LB6, represented by several hand bones, a scapula (shoulder blade) and a mandible, LB7, a tiny, but adult, bone from the thumb, LB8, another tibia, and LB9, a femur" (Groves, 2007, p. 125). Interestingly, all of the newer remains discovered belonged to individuals with a shorter stature than LB1 (Groves, 2007). The tiny stature of more recently discovered *H. floresiensis* fossils further establish that the pathological modern human theories are unlikely. Furthermore, both LB1 and LB6 show the same mandibular morphology outside the range of the modern human (Groves, 2007). It was also determined that *H. floresiensis* dated past 12 000 years, close to 60,000 to 100,000 years ago (Sutikna, 2016). Modern humans only reached Australia and Southeast Asia about 50,000 years ago, making it more likely that *H. floresiensis* are a distinct human species rather than descendants of *H. sapiens* (Sutikna, 2016).

Conclusion

Discovering a new human species can bring about a great deal of skepticism and criticism due to its evolutionary implications. This was the case with the discovery of *H. floresiensis* fossils, causing tremendous debate as to whether or not it is a distinct human species. This debate stems from its unique characters: its very small stature accompanied with a small brain. The main argument that *H. floresiensis* is a pathological or diseased modern human has been disproven, and it is infrequently referenced in the modern literature. With the additional fossil discoveries of LB2-LB9, it is evident that the small stature of LB1 is not an exception. The entire population shares a small stature, making it unlikely that it was due to disease. When the entire morphology of the *H. floresiensis* skeleton was considered, it was determined to be more similar to that of apes rather than the modern human (Baab, 2012). This led to the conclusion that *H. floresiensis* is more likely a descendant of another human species known as *Homo habilis*, which is older than *H. erectus* (Groves, 2011). Although there has been great scientific debate over the years, *H. floresiensis* is now widely recognized as a distinct human species.

Chapter 6: Ethical Concerns in Human Genetic Research (Shahreen Rahman)

Introduction to Genetic and Evolutionary Research

Genetic and evolutionary research has been continually built upon since its initial discovery. and helps expand our understanding of genetic history (Randolph, 2008). Genetic research includes exploring gene homology, and examining data from a species, as well as the ways they have adapted from processes such as natural selection (Randolph, 2008). Homology can be defined as the similarities that arise due to common ancestry (Campbell, 1988). Natural selection is the process of beneficial traits being reproduced due to the fact that it is useful to survival, or the concept of "survival of the fittest" (Gregory, 2009). This can help look into an individual's susceptibility to certain diseases and predict their future health outcomes (Australian Law Reform Commission and Australian Health Ethics Committee, 2001).

While genetic and evolutionary research is undeniably valuable, there are many ethical considerations that need to be addressed before conducting research. If they are not acknowledged and incorporated, the methodology and results of the research can pose risks, be misconstrued and applied in ways that negatively impact the subjects and the groups the research is being done. This is also why Research Ethics Boards and Institutional Review Boards exist to protect the rights of subjects (Page and Nyeboer, 2017).

Ethical Guidelines and Framework in Research

Research ethics constitutes acting with integrity, honesty, compassion, and ensuring the safety of everyone participating in the research. Research ethics serves as a set of guidelines for anyone performing research, ensuring that researchers act properly and deliver reliable results. One of the most important aspects of research ethics is to observe the framework, which identifies four key moral concepts (Collier and Haliburton, 2015).

The four basic principles begin with Respect for persons, which ensures that a person has autonomy but still follows procedures (Collier & Haliburton, 2015). The next principle is non maleficence, which protects participants from harm (Collier & Haliburton, 2015). The following principle is benefiance, the study should benefit people, and lastly justice, which is that those who are involved in the research should benefit in some way and the research should be applicable to a diverse number of groups (Collier & Haliburton, 2015).

The first and second principles ensure that harmful outcomes are considered against potential benefits and mitigated accordingly (Collier & Haliburton, 2015). The Justice principle protects participants from harm and ensures that research produced is productive and helpful to society. Secondly, this part of the framework maintains that researchers act professionally and uphold a standard of integrity and care. These principles uphold all participants need to be informed about the research, give consent, and be protected in terms of their health and privacy (Collier & Haliburton, 2015). The third principle would be used to determine that the research has applications that have a scope of benefit. The final principle ensures that all participants are treated equally and compensated for their participation, and that they benefit from the research in some way (Collier and Haliburton, 2015). When conducting any type of research, these four aspects of the framework are absolutely critical in ensuring that all those involved benefit from the process of and outcomes of research.

Vulnerable Populations

Ethical principles become particularly important when conducting research involving vulnerable populations. Vulnerable populations are groups who have either structural or systemic vulnerability to those in power, and it is those in power who are conduct the research (Collier & Haliburton, 2015). Vulnerable groups may choose to participate in the research due to coercion such as the research helping the vulnerable individual receive a right or freedom they would not otherwise be able to attain easily (Collier & Haliburton, 2015).

There are four elements which define vulnerable populations; the first is the institutionalized disparity of power (Collier & Haliburton, 2015). Institutionalized disparity of power means those who are most vulnerable are people at the bottom of a structure or system (Collier & Haliburton, 2015). The second element is that vulnerable populations have an inability to respect their best interests (Collier and Haliburton, 2015). For example, if one were in a vulnerable situation, they likely had to endure negativity in order to function when in normal cases would not have to endure such conditions, such as experiencing discrimination (Collier and Haliburton, 2015). The third element of vulnerability is whether the population has to consider punishment or rewards before participating (Collier & Haliburton,

2015). This means that those involved are not participating out of free will but rather out of fear of consequences or the enticement of a benefit. The final element is susceptibility, which addresses how much power others have to influence or motivate someone to do something or take over aspects of life (Collier & Haliburton, 2015). This could apply to the situation of a prisoner who is very susceptible as they have little to no power, so others in power can enlist fear to coerce them into actions or participation.

A History of Misconstrued Ethical Applications of Human Genetic and Evolutionary Research

Within human biology and medicine, new evolutionary questions about topics such as host–parasite–symbiont relationships and pathogen evolution interactions are becoming more commonly discussed (Gluckman et al, 2011). Evolutionary medicine is a new field that has emerged from the intersections of evolutionary biology, clinical medicine, and experimental biomedical disciplines; it explores evolutionary explanations of disease vulnerability (Gluckman et al, 2011). Human evolutionary research has become a major topic in health applications and thus it is important to consider the ethical implications that arise.

As per the beneficence principle, the research should be beneficial to the subject participating in some form (Collier & Haliburton, 2015). However there have been cases where rather than benefiting humans, the research instead caused long standing negative impacts toward vulnerable populations. The primary example here is the use of evolutionary research in justifying racism (Graves, 2019).

Racism can be defined along with prejudice and discrimination which are defined as biased feelings and unequal treatment respectively (Salter et al, 2017). It is thought that "[t]he Western concept of race, although socially constructed, was always rooted in 'biological' features; that is, the characteristics used to classify individuals into races were always assumed to have a basis nontrivially rooted in human biology" (Graves, 2001, p. 37). One of the leading arguments that many use to justify racism stems from the question posed by evolutionary biologist J.B.S. Haldane, "are the biological differences between human groups comparable with those between groups of domestic animals such as greyhounds and bulldogs…?" (Norton et al, 2019, p. 1). As dogs were a focal point of Darwin's research on evolution and he explored comparing human races with dog breeds in *The Descent of Man* (1871), many look to attempt to conform their theories to those of Darwin's (Norton et al, 2019). Darwin himself also believed that white races were more progressive than other races (Rose, 2009). This analogy isn't just an academic evolutionary question but also one that questions the basis of racism as a social construct and tries to misconstrue science to justify it. In this analogy, purebred dogs allude to the concept of being 'pure' (Norton et al, 2019) and this is used by white supremacists to confirm their notion of 'human racial purity' (Norton et al, 2019). An example of this is in the interview from 2016 conducted by Mother Jones with a white supremacist, who used this very arguement of comparing the human race to dog breeds (Norton et al, 2019).

Even recent publications have been used to derive racist conclusions with one such case being the proposed evidence for natural selection acting on two genes found primarily in Eurasian populations, however not in West African populations (Vitti et al, 2012). Since these genes are linked to microcephaly and non-progressive mental retardation, the researchers speculated that these genes could affect brain size and may have been impacted by recent natural selection (Vitti et al, 2012). This particular hypothesis led to the forming of racist conclusions such as one race holding superiority.

Darwin also used his evolutionary research particularly on sexual selection to define man as "more courageous, pugnacious and energetic than woman [with] a more inventive genius. His brain is absolutely larger [while] the formation of her skull is said to be intermediate between the child and the man" (Darwin, 1871, p. 316). His take on gender can also be misconstrued to promote a gender hierarchy, and does not provide all the information as it is most commonly males that evolve to meet the female sexual selection choices (Rose, 2009).

Another example of the misuse of evolutionary research to target vulnerable populations is the eugenics movement, which is based on the studies of Francis Galton who coined the term eugenics meaning "well born" (Norrgard, 2009). The term was created after he discovered that desired traits from English families such as intelligence could be passed on successfully at a rate of 20% (Norrgard, 2009). This led to the eugenics movement in the 1900s where eugenists believed that they could control human mating, based on researching an individual's genetic information and data (Norrgard, 2009). The pedigrees collected were also affected by prejudices and biased interpretations (Norrgard, 2009). Political leaders, for example Alexander Graham Bell and Winston Churchill, agreed that the improvement of the human race based on the information provided by pedigrees and evolutionary research should be promoted from a policy level through government intervention (Farber, 2008). This idea was first expressed through the endorsment of selective breeding, but eventually led to state-sponsored discrimination, forced sterilisation, and genocide (Farber, 2008).

During the immigrant boom in America in the late 1800s and early 1900s, eugenics, particularly by assigning immigrants with traits such as mental illness and criminality, prevented the government from addressing the inefficient social resources in place (Norrgard, 2009). After the discreditation of eugenics in the United States, Farber (2008) quoted Jonathan Marks (1996),

> "Genetics was corrupted in the 1920s by the confusion of folk knowledge with scientific inference. For whatever reasons, outsiders who recognized it were shunned, and insiders were, as they say, a day late and a dollar short. The fairly obvious lesson to be learned is that where science appears to validate folk beliefs, it needs to be subjected to considerably higher standards of scrutiny than ordinary science (Farber, 2008, p. 245)"

This sentiment however did not stop eugenics from being used. In fact in the 1930s, eugenics was used to sterilize 400,000 Germans without consent as they were deemed to have undesirable traits (Norrgard, 2009).

Currently, the concept of evolution can still be misunderstood, and thought by some as the progress to a particular 'fixed ideal' (Vitti et al, 2012). This misinterpretation gives the public the reading that evolution is the "preservation of favoured races in the struggle for life" as said by Henry Spencer quoted by Vitti et al (2012, p. 4).

Conclusion : Considering the Framework when Conducting Research

As human technologies progress and more discoveries are made on the human genome, three important ethical issues will need to be addressed. The first of these is that more research will be available on the human genome based on individuals and populations. Who should decide how research is applied and disseminated? The second consideration is the ability to manipulate human genotypes and phenotypes based on the research (Murray, 1991). Lastly the discoveries made on desired traits that are valuable politically can change how we perceive ourselves and social institutions (Murray, 1991).

It is evident that due to how research is disseminated, communication of results needs to be considered as the use of media can lead to sensationalization of information and the greater population forming their own interpretations (Vitti et al, 2012). Researchers must exercise caution and propose questions as to what can be considered evidence to ensure that the methodology for the research is rigorous (Vitti et al, 2012).

In some cases the application of evolutionary research can lead to long lasting and irreversibly devastating effects, as seen by the use of evolutionary research to justify racism and the eugenics movement. Murray (1991) highlighted a lack of ethical practice leading to the manipulation of the human genotype. Research ethics boards could have prevented the outcomes of the eugenics movement as well as the use of evolutionary research to justify racism. Due to the implications that could arise, it is of utmost importance to consider the framework and ethical guidelines when conducting and applying evolutionary human research as it can be misused to isolate and discriminate against vulnerable populations.

Chapter 7: Comparative Neuroanatomy of *H. floresiensis* and the Modern Human (Omar Hadi)

Homo floresiensis is one of the most recently discovered human species. They lived close to 100,000 to 50,000 years ago (Martin et al., 2004) Nicknamed 'the hobbit', *Homo floresiensis* stood at approximately three feet and six inches (106 cm) and weighed approximately thirty kilograms (Martin et al., 2004*)*. They had small brains, relatively large feet and teeth, a shrugged-forward posture and disproportionate limbs. However, despite their very small brains and body size, researchers believe that *H. floresiensis* made tools from stone, hunted large rodents and small elephants, and they may have even used fire (Martin et al., 2004). Researchers also believe that *H. floresiensis* underwent insular dwarfism which may have caused the reduction of body size and brain size. Insular dwarfism, also known as island dwarfism, occurs when a population is confined to a small environment, usually islands. When a population is confined to such environments and food levels gradually decrease, individuals with the least energy requirements survive. Over many generations, the overall size of the species decreases to aid survival, as smaller individuals require less food and resources which gives them an evolutionary advantage.

The evolution of the human brain has had a significant impact on the way of life of modern humans. Modern human brains are about three times bigger than our *Homo floresiensis* ancestors. Using modern high-resolution CT scans, scientists have estimated that the size of the *Homo floresiensis* brain is approximately 426 cm^3 (Than et al., 2021). In comparison, the size of the modern *Homo sapien* brain is approximately 1400-1500 cm^3 (Than et al., 2021). This indicates compared to *Homo floresiensis* brains, modern human brains are approximately three times larger in terms of endocranial volume (ECV), a measure of the volume of the interior of the skull (Colby et al., 2021). To further compare the dimensions of the *Homo sapien* and *Homo floresiensis* brains, researchers used three-dimensional computed tomographic reconstructions of both species (Falk et al., 2007). These reconstructions recreated important details of the brain's anatomy such as vessels, sinuses, shape and endocranial volume (Falk et al., 2005). The compared measurements included length, width, and breadth (Falk et al., 2005). These reconstructions showed that *Homo sapien* brains are 49 cm longer and 40.6 cm wider than *Homo floresiensis* brains, which may explain the advanced behavior of *Homo sapiens* (Falk et al., 2005) .

After the discovery of the *H. floresiensis* remains in 2003, researchers were surprised by the very small size of the cranium, which are all the bones that formulate the human skull. In order to compare brain sizes of both *H. sapiens* and *H. floresiensis*, researchers used craniometric ratios (measurements of the dimensions of the skull) to differentiate between microcephalic and normocephalic individuals (Vannucci et al., 2011). Microcephaly is a birth defect where the cranium and the brain are smaller as compared to normal individuals of the species (Vannucci et al., 2011). Normocephalic translates to 'normal head'.

The two major craniometric ratios used for the comparison between brain sizes are the cerebellar protrusion and the relative frontal breadth ratios (Falk et al., 2005). The cerebellar protrusion is calculated by dividing the cerebellar pole (projected frontal pole) by the cerebral length and the relative frontal breadth is calculated by dividing the frontal breadth by the cerebellar width (Vannucci et al., 2011). To compare these ratios, scientists used normocephalic individuals as controls and calculated the cerebellar protrusion and the relative frontal breadth for *H. floresiensis*. These ratios were calculated to be 0.965 and 1.04 for cerebellar protrusion and relative frontal breadth respectively (Falk et al., 2005). Both ratios calculated are outside the normocephalic range and therefore it was concluded that *H. floresiensis brains* were brachycephalic, where brachycephalic refers to a 'shortened-head' (Vannucci et al., 2011). Many hypotheses have been proposed to explain why the endocranial volume of *H. floresiensis* is drastically reduced. Among these hypotheses, some scientists have said that *H. floresiensis* is modern *H. sapiens* with microcephaly (Vannucci et al., 2011). It was also observed that the cranium of *H. floresiensis* is similar in shape to a modern day human with microcephaly. This hypothesis has been disproven after the calculation and comparison of the cerebellar protrusion and the relative frontal breadth ratios (Falk et al., 2005). Scientists have shown that *H. floresiensis* craniums fall outside the range of variation for human microcephalic skulls (Hershkovitz et al., 2007).

Laron syndrome is a condition in which the body cannot effectively use growth hormone (GH) and usually results in developmental defects and a short stature (Hershkovitz et al., 2007). Laron syndrome is usually caused by mutations to the

GHR gene and is an autosomal recessive disorder (Hershkovitz et al., 2007). Adult males diagnosed with Laron syndrome typically reach a maximum height of four and a half feet and adult females typically reach a maximum of four feet in height (Hershkovitz et al., 2007). A hypothesis was postulated that the *H. floresiensis* remains discovered in 2003 are *H. sapien* remains of individuals that suffered from Laron syndrome (Hershkovitz et al., 2007).

Although thorough DNA analysis would be required to fully confirm this theory, it seems unlikely because the extremely small endocranial volume of *H. floresiensis* is much smaller than that observed in modern patients with Laron syndrome. Another hypothesis states that *H. floresiensis* are endemic cretins (Falk et al., 2005). Endemic cretinism, also known as congenital iodine deficiency syndrome, is a medical condition caused by insufficient thyroid hormone causing hypothyroidism (Hershkovitz et al., 2007). It is usually caused by a lack of iodine during pregnancy. Hypothyroidism can cause an individual to develop cretinism which is characterized by small bodies and reduced brain size. Researchers have argued that *H. floresiensis* has the basic features of cretinism such as an enlarged pituitary fossa (Falk et al., 2005). The pituitary fossa is a small depression in the sphenoid bone of the human skull and is located directly under the nose. However, when the *H. floresiensis* pituitary fossa was scanned, it showed no enlargements (Falk et al., 2005). *H. floresiensis* cranial bones have also been compared to ten individuals that had cretinism and limited similarities were found (Falk et al., 2005). Furthermore, the *H. floresiensis* neurocranium is considerably different from humans with either Laron syndrome or endemic cretinism (Falk et al., 2005). This difference suggests that endemic cretinism or Laron syndrome are unlikely causes of the *H.*

floresiensis neurocranium (Falk et al., 2007). A third hypothesis was postulated that questioned the remains of the H. floresiensis individuals that were discovered. A 2014 study claimed that the remains found were simply modern humans with Down syndrome (Falk et al., 2005). Down syndrome is a genetic condition in which an individual is born with an extra copy of chromosome 21. However, it is unlikely that the *H. floresiensis* remains were humans with Down syndrome because the remains found had characteristics not found in individuals with Down syndrome. These characteristics include the absence of a chin and in general, their height was taller than individuals with Down syndrome (Falk et al., 2005).

The main difference between *H. floresiensis* and *H. sapiens* is that *H. floresiensis* are characterized by their small size and small endocranial volume (Martin et al., 2004). However, less obvious features include the shape and formation of the teeth, a lesser torsion around the waist area and the absence of a chin (Martin et al., 2004). The *H. floresiensis* brain is approximately one half the size of its ancestor *Homo erectus* (*H. erectus*) and one third the size of the *H. sapien* brain. After establishing that *H. floresiensis* are in fact part of the *Homo* genus, researchers set out to study their anatomy and psychology and its implications on the evolutionary history of humankind. It is believed that *H. floresiensis* were an advanced species, they made tools for hunting using rocks and sticks, they hunted animals ranging in size from large rodents to small elephants, and they even knew how to use fire (Martin et al., 2004). The extremely small brain size of *H. floresiensis* does not appear to have affected their mental capacities. This is primarily due to a specific area of the brain called Brodmann's area 10 located on the dorsolateral prefrontal cortex. Brodmann's area 10 (BA10) is the most anterior portion of the prefrontal cortex in the human brain and is found in both *H. floresiensis* and *H. sapiens*. Cranial measurements of *H. floresiensis* brains shows an enlargement of the frontal polar region of the brain which may indicate an enlarged BA10 area (Falk et al., 2005). An enlarged BA10 area may indicate an advanced level of cognition, BA10 in *H. floresiensis* shows bilaterally expanded gyri but appears normal; this is further evidence that they were not microcephalic *H. sapiens* (Falk et al., 2007). The prefrontal cortex (PFC) is the cerebral cortex that covers the superior region of the frontal lobe (Walters et al., 2020). The prefrontal cortex contains all the Brodmann's areas ranging from BA8 to BA47. Although the function of BA10 is not entirely known, it is believed that BA10 is the center for cognitive branching and cognition (Chanhine et al., 2015). Cognitive branching is the ability to perform tasks related to one goal while keeping in memory information related to a second goal (Chahine et al., 2015). The prefrontal cortex is the main site of executive functions such as cognition, short term memory, planning and decision making(Walters et al., 2020). The small size of the *H. floresiensis* brain does not appear to have had an effect on the mental capabilities of the species as BA10 is approximately the same size in both *H. floresiensis* and modern day humans (Falk et al., 2005). Furthermore, *H. sapien* frontal lobes are significantly larger than *H. floresiensis* frontal lobes. The frontal lobes are responsible for voluntary movements, speech and higher executive function. This may explain why modern day humans are more cognitively complex than any of their ancestors.

Although the basic brain anatomy is similar, the cranium of *H. floresiensis* is very different from the cranium of *H. sapiens,* and is very similar to the cranium of the *H. erectus*. The brain anatomy of *H. floresiensis* strongly indicates a phylogenetic relationship with its ancestor *H. erectus* (Falk et al., 2005). However, after careful analysis using cranial endocasts (internal cast of a hollow object), some characteristics of *H. floresiensis* appear to be derived traits which have appeared due to random mutations in the genome. The orientation and gyrification of the lateral prefrontal cortex relative to the temporal poles in the *H. floresiensis* appear to be derived as they are not present in the anatomy of the *H. erectus* (Falk et al., 2005).

H. floresiensis were an advanced species that lived about one hundred thousand years ago who used stone tools to hunt large animals and used fire to survive (Martin et al., 2004). They likely arose due to a phenomenon known as insular dwarfism which is an evolutionary process where a population becomes isolated in an area with scarce resources. Many researchers have questioned the evolutionary origins of the *H. floresiensis* fossils discovered in 2003. Many hypotheses have been proposed criticizing the discovery. Three of these hypotheses claim that the *H. floresiensis* remains found were either modern humans with Laron syndrome, endemic cretinism or Down syndrome (Falk et al., 2005). These hypotheses have been disproven by new technological advances Neurocranial measurements such as the cerebellar protrusion and the relative frontal breadth show that size of the brain of *H. floresiensis* falls outside the range of variation for the physiopathologies discussed , further showing that *H. floresiensis* were indeed a newly discovered species of the *Homo* genus. *H. floresiensis* is characterized by a small body as well as a very small brain and endocranial volume. Their brains are one third the size of modern humans and have a size of approximately 417 cm^3; but the small size of the *H. floresiensis* brain did not have a significant effect on their cognition and mental capabilities, owing to a fully sized and functional Brodmann's area 10 (BA10). Modern humans may be more complex due to a much larger prefrontal cortex which is the center of many executive functions.

Chapter 8: Comparative Anatomy Between *H. floresiensis* and *H. erectus* (Shannon Lin)

Introduction

Commonly referred to as 'the Hobbit,' the stature of *H. florensis* has most notably been characterized as uniquely small-bodied, short-legged, and even small-brained (Gruss & Schmitt, 2015). In addition to evaluating the significance of its size and its succeeding scientific discoveries, it is also vital to assess how the anatomy of *H. floresiensis* compares to those of other human ancestors that were also introduced during the early Pleistocene, which occurred just over two million years ago (Gruss & Schmitt, 2015; Larson et al., 2007). This chapter will compare and contrast various anatomical elements of *H. floresiensis* and its closest evolutionary counterpart within this period, *H. erectus*. As previous fossil evidence of the genus *Homo* has suggested a likelihood of evolutionary patterns over time, the study of these remains are important as they provide an indication of the anatomical conditions that concern selective priorities during different phases of human evolution (Gruss & Schmitt, 2015). From this perspective, it is of utmost importance to evaluate the anatomical conditions of *H. floresiensis* and compare them with other human ancestors, as it can pinpoint its origins and tell the story of their time on Earth.

H. floresiensis and *H. erectus*

Various primitive features of *H. floresiensis* have been put at the forefront of research that establish anatomical similarities and differences with the features of *H. erectus* (Lieberman, 2009). Believed to be closely attributed to the ancestry of its preceding species, *H. erectus*, this indicates the possibility for *H. floresiensis* to evolve via insular dwarfism from Javanese *H. erectus* or from early, non-Asian *H. erectus* (Churchill & Vansickle, 2017; Lieberman, 2009). Common to islands such as Flores in Indonesia, insular dwarfing refers to an evolutionary process where large species become small by undergoing intense selection (Kühne, 2019; Lieberman, 2009). The resulting primitive features (features that are found in most australopiths) include *H. floresiensis* to be significantly shorter in an absolute sense, which can be exhibited through their relatively small joints and limbs overall (Churchill & Vansickle, 2017; Larson et al., 2009; Lieberman, 2009). Examples of such features are described in the article by Baab, McNulty, and Harvati (2013), who found *H. floresiensis* to resemble anatomical structures found in hypothyroidism and Laron syndrome, an endocrine disorder, but they were found to not derive from such conditions. These are shown through a smaller skull, absent frontal sinuses, delayed development of the clavicle and scapula, high humerofemoral index, thick humeral shafts, a reduced humeral torsion, and defective growth hormones that lead to stunted growth caused by primarily foreshortened legs (Baab et al., 2013).

Skull and Upper Limb Comparisons in *H. floresiensis* and *H. erectus*

At the skull, Lieberman (2009) described *H. floresiensis* to have a vertical face, no snout, and most teeth that closely resembled those of *H. erectus*. This upholds previous claims based on craniofacial features that have attributed the ancestry of *H. floresiensis to H. erectus* (Antón, 2012; Baab et al., 2013). A summary of their craniofacial similarities include having "a low cranial vault, thick cranial bones, a frontal keel, mastoid fissure, and a fissure between the tympanic plate and entoglenoid pyramid" (Baab et al., 2013, p. 1). While these features in *H. floresiensis* and *H. erectus* are synonymous, Antón (2012) argued that their physical differences, to be discussed further, "suggest underlying adaptive shifts at the origin of the genus *Homo*" (p. 279).

Below the skull finds the shoulders and clavicles. While *H. floresiensis* has a relatively short clavicle, protracted scapulars, and an anteriorly facing glenoid fossa (facing the front), *H. erectus* instead has a longer clavicle, inferiorly rotated and ruman-like scapulars, and a laterally facing glenoid fossa (Larson et al., 2007; Roach & Richmond, 2015). Precisely, the scapulars in *H. floresiensis* are generally perceived to be human-like but much smaller in size, while the clavicles are found to be short and very curved (Larson et al., 2007; Lieberman, 2009). In Larson, Jungers, Morwood, Sutikna, Jatmiko, Saptomo, Due, and Djubiantono's (2007) study on the evolution of the hominin shoulder and *H. floresiensis*, the authors suggested that this configurational shift toward a shorter clavicle represented "a transitional stage in pectoral girdle evolution in the human lineage" (p. 718). These structural differences enabled the ideal capacity for *H. erectus* to engage in high speed throwing and running that was necessary for survival (Larson et al., 2007; Roach & Richmond, 2015). When combined with a tall, mobile waist, and low humeral torsion, this combination in upper limb morphology created sufficient conditions for proficient throwing abilities that were most prevalent in *H. erectus* (Roach et al. 2013, as cited in Roach & Richmond, 2015). This selective force for clavicular elongation in *H. erectus* through running can be best explained by the requirement for "shoulder and upper body rotation to counteract the destabilizing torque created by the oppositely moving lower limbs" (Larson et al., 2007, p. 728).

Lieberman (2009) states that the straight humerus bone within the upper limbs of *H. floresiensis* "[lack] the normal degree of twisting between the shoulder and the elbow," which results in a minimal amount of humeral torsion, which is also prevalent in *H. erectus* as previously mentioned (Larson et al., 2007; p. 42; Roach et al. 2013, as cited in Roach & Richmond, 2015). These findings therefore raise further questions regarding the actual course of hominin shoulder evolution during the early Pleistocene (Larson et al., 2007). With wrists comparable to those of apes, this also demonstrates the compacted nature of the *H. floresiensis* bodies which were used for tool making, as well as for hunting dwarfed elephants and giant varanid lizards on the island (Gruss & Schmitt, 2015; Lieberman, 2009).

Pelvic Comparisons in *H. floresiensis* and *H. erectus*

Similar to the findings on each species' shoulder composition and how they are impacted by selective forces such as throwing, pelvic morphology remains to be a prominent component of comparative research (Gruss & Schmitt, 2015). This is because it provides essential insights on changes in locomotor behaviours amongst various species (Gruss & Schmitt, 2015). Considering how the pelvis is the "primitive overall form established in early hominins," this distinction is clearly shown in the research that found significant size differences in the pelvic anatomy of *H. floresiensis* and *H. erectus* (Gruss & Schmitt, 2015, p. 7). In particular, the pelvis of *H. erectus* is mediolaterally broad and anteroposteriorly narrow, with a platypelloid-shaped birth canal and laterally flaring iliac blades that resemble butterfly wings at the top (Gruss & Schmitt, 2015). This widened and rounded structure is emphasized in Churchill and Vansickle (2017), where the primitive features of the *H. erectus* pelvis include laterally flaring iliac blades that are found only within this species and in most australopiths.

Other characteristics that are exclusive to *H. erectus* and not found in humans include a pronounced iliac buttressing and a laterally-oriented ischial tuberosity, meaning that the trajectory of the structure follows outwards from the centre of the body (Churchill & Vansickle, 2017). Features of the *H. erectus* pelvis that resemble those of modern, living humans', rather, include a more pronounced S-shaped sigmoid curvature of the iliac blade with a narrow tubero acetabular sulcus below the iliac region (Churchill & Vansickle, 2017). It was concluded by Gruss and Schmitt (2015) that while the pelvis of *H. erectus* is broad when compared with modern humans, it is actually narrower relative to body height than the hominin remains found from the australopithecines, its preceding species. When comparing the australopithecines and *H. floresiensis*, they both acquire small joint surfaces on the ossa coxae, an inclusive term that refers to the pelvic bones in their entirety (Churchill and Vansickle, 2017). These conclusions therefore suggest further inquiry between the

Homo lineage between *H. floresiensis* and *H. erectus* (Churchill & Vansickle, 2017).

Within the small-bodied *H. floresiensis*, its similarities and differences in pelvic morphology are found to "represent the evolutionary loss of the derived pelvic and proximal femoral features found in both early African and later Asian *H. erectus*." (Churchill & Vansickle, 2017, p. 974). Similar to *H. erectus*, its iliac blades are markedly laterally-flared (Churchill & Vansickle, 2017). Through *H. floresiensis* remains studied by Jungers, Larson, Harcourt-Smith, Morwood, Sutikna, Awe, and Djubiantono (2009a), this particular degree of iliac flaring is strongly associated with the ossa coxae that are found in australopithecines as opposed to modern humans and Pleistocene hominins (McHenry, 1975; Stern and Susman, 1983; Arsuaga et al., 1999; Lovejoy, 2005, as cited in Jungers et al., 2009a). Opposing features in *H. floresiensis* when compared with *H. erectus* involve having weakly developed and anteriorly positioned acetabulo-cristal buttresses with small cristal tubercles, and a wide groove for the obturator internus muscle (tubero acetabular sulcus) on the ischium (Brown et al., 2004 and Jungers et al., 2009, as cited in Churchill & Vansickle, 2017). In short, the pelvic shape overall is "consistent with the expectations for a small-bodied hominin with a very small cranial capacity" (Jungers et al., 2009a, p. 541). This characteristic of a smaller pelvic shape is of particular importance within the context of comparing the pelvic morphology with that of an adult female *H. erectus*. Founded in the Acheulean-bearing deposits of the upper Busidima Formation at Gona, Ethiopia, the Gona pelvis introduced a notable discovery in the research of *H. erectus* and human evolution overall (Churchill & Vansickle, 2017).

Associated with an adult female *H. erectus*, the Gona pelvis is broad and consists of a capacious birth canal, which results in a rounder, gynecoid pelvic inlet (Churchill & Vansickle, 2017, p. 970; Gruss & Schmitt, 2015). Citing Simpson and colleagues (2008), both Churchill and Vansickle (2017) and Gruss and Schmitt (2015) emphasized that this large pelvic outlet that shaped the birth canal occurred due to adaptations for birthing babies with larger brains. This evidence suggests that pelvic morphology had changed to adapt to greater encephalization in early *Homo*, which refers to *H. erectus* within this context (Churchill & Vansickle, 2017). Furthermore, this remarkable discovery of the Gona pelvis demonstrates the postcranial diversity that existed among Pleistocene hominins (Gruss & Schmitt, 2015). It was also suggested that a narrower pelvis is advantageous in terms of locomotor efficiency with an increased ligamentus stabilization of the pelvic joints (Churchill & Vansickle, 2017; Gruss & Schmitt, 2015; Lieberman, 2009). A narrower pelvis also brings forth increased thermoregulation in hot climates as opposed to a wider pelvis, where these differences in pelvis sizing overall justifies the act of insular dwarfism that *H. floresiensis* underwent by natural selection within a warm climate (Churchill & Vansickle, 2017; Gruss & Schmitt, 2015; Lieberman, 2009).

Lower Limb Comparisons in *H. floresiensis* and *H. erectus*

Below the bony pelvis finds a combination of lower limbs including the femora, tibiae, metatarsals, and pedal phalanges, which are described to be short in length and relative to the small composition of *H. floresiensis* overall (Jungers et al., 2009a; Jungers et al., 2009b; Lieberman, 2009). These essentially correspond to the thigh bones, shin bones, long bones in the foot, and short bones in the toes. Found on the shaft's posterior surface (back side) of *H.floresiensis'* femora, Jungers and colleagues (2009a) identified a prominent and well-defined line that traces the upper lateral branch of the linea aspera. Along the linea aspera, the authors found that the gluteal tuberosity is significantly sunken in, which creates a small and concave region at the top of the shaft. This combination of features is described to correlate with an increased shaft robusticity, which enables a stronger limb structure in *H. floresiensis* (Jungers et al., 2009a). Most importantly, the greater trochanter on the *H. floresiensis* sample exhibits evidence of erosion through prominent lateral apophysis (Jungers et al., 2009a). This is a significant characteristic exclusive to the femora of the genus *Homo*, and this distinguishes it from the lateral trochanteric surfaces that are typically found in australopithecines (Harmon, 2008, as cited in Jungers et al., 2009a). Another similarity was drawn by both Jungers and colleagues (2009a) and Antón (2012), who found that *H. floresiensis* and *H. erectus* shared an anteroposteriorly compressed femoral neck, located above the tibia.

Within the tibia of *H. floresiensis*, Jungers et al. (2009a) drew the observation that there is a very slight curvature along the long axis. Quoting the findings by Brown and colleagues (2014), the cross-section at its midshaft is oval in shape, and it lacks a sharp anterior border or crest (Jungers et al., 2009a). Jungers and colleagues (2009a) conclude that "this functional inference lacks biomechanical substance" (as cited in Jungers et al., 2009a, p. 545). The authors further described that the tibiae are 'robust' in nature, exemplifying that its transverse diameters are considered to be large for the lengths of the bones. In relation to its ancestral lineage and likelihood of undergoing intense transformations through insular dwarfism, Jacob et al. (2006) recall that the cortical bone in the tibiae of the *H. floresiensis* sample, in particular, is pathologically thin and occurred as a result of long bone "over-tibulation" and "disordered growth" (cited in Jungers et al., 2009a, p. 545).

One primitive feature of the *H. floresiensis* foot includes toes that are long, curved, and robust, with the exception of a short big toe (Lieberman, 2009). The navicular bone, which is considered to be a crucial bone, provides a weight-bearing process that acts "like the keystone at the top of the inside of the human arch" (Lieberman, 2009, p. 41). This weight-bearing navicular tuberosity, combined with a short hallux, long lateral rays, and the lack of a well-defined medial longitudinal arch in the foot of *H. floresiensis* limits the recovery of stored elastic energy, which hinders the mass-spring mechanics that are activated when running (Jungers et al., 2009b). In *H. floresiensis*, it was most notably found that a locking mechanism found in the middle of the foot stiffened itself as the calf muscle raised the heel off the ground, which allocated well for effective walking but ill-designed for high-speed or efficient endurance running (Jungers et al., 2009b; Lieberman, 2009).

Conversely, Gruss and Schmitt (2015) noted that the longer lower limbs and higher crural indices obtained by *H. erectus* would have enabled them to "walk farther at a lower energetic cost" (p. 7). This also hinted that "endurance running may have been a component of the hominin locomotor repertoire for the first time in *H. erectus*" (Gruss & Schmitt, 2015, p. 7). This could be demonstrated anatomically by analyzing the overall characteristics of the lower limbs of *H. floresiensis*, where they were found to have enlarged articular surface areas of the long bones with a thick cortical bone in the lower limb (Weidenreich, 1941, as cited in Antón, 2012). By combining the kinematic and structural properties of the pelvis, lower limb bones, and feet of *H. floresiensis*, it can be argued that without a doubt, it led toward an adaptation to bipedalism (Gruss & Schmitt, 2015; Jungers et al., 2009b). To be further discussed in Chapter 11, it is important to recognize the effects of selective forces that have impacted various anatomical conditions, such as bipedalism and the lower-body biomechanics, that were founded and developed throughout the different stages of human evolution.

Conclusion

Early findings of insular dwarfism have indeed justified the small and outrageously compacted nature of *H. floresiensis* in all senses (Baab, McNulty, and Harvati, 2013; Churchill & Vansickle, 2017; Gruss & Schmitt, 2015; Kühne, 2019; Larson et al., 2009; Lieberman, 2009). While it has been previously hinted that the anatomical features of *H. floresiensis* have been derived from the *H. erectus* lineage, these comparisons from a selection of literature and studies have established the unique ways both species differ and relate with each other (Churchill & Vansickle, 2017; Lieberman, 2009). By finding similarities in the facial, femoral, and humoral structures, and differences in pelvic and lower limb morphology, the evidence demonstrates how each species-although close in origin and anatomical forms-have adapted according to varying functionalities and biogeographical factors (Antón, 2012; Churchill & Vansickle, 2017; Gruss & Schmitt, 2015; Jungers et al., 2009a; Jungers et al., 2009b; Lieberman,

2009; Larson et al., 2007; Roach et al. 2013, as cited in Roach & Richmond, 2015). Exemplified through the need to hunt and walk in *H. floresiensis* as opposed to *H. erectus*, who were well structured to birth babies with a larger cranial capacity, run, and throw, these comparisons have illustrated how each species within the same *Homo* genus have prioritized their capabilities dependent on their anatomical composition (Simpson et al., 2008, as cited in Churchill & Vansickle, 2017; Simpson et al., 2008, as cited in Gruss and Schmitt, 2015; Jungers et al., 2009b; Lieberman, 2009; Larson et al., 2007; Lieberman, 2009; Roach & Richmond, 2015). This comparison between *H. floresiensis* and *H. erectus* have indeed shown that comparative anatomical studies for various hominins are essential, and are encouraged to maintain the integrity of anthropological studies that aim to develop a comprehensive understanding of human evolution.

Chapter 9: The Abnormal Brain Size of *H. floresiensis* and its Implications (Michaela Dowling)

The Importance of Examining our Neurological Evolution

To articulate the complexities of human nature, both historically and as observed within societies today, one must not neglect understanding the foundation of all human behavior: the brain. Since the beginning of life, natural selection has proven to be the driving force behind evolution (Brown, 2001). In particular, divergent evolution has allowed organisms to become better suited to their ecological niches, in turn, diversifying the species that inhabit the planet today (Brown, 2001). It would be an overstep to declare the *Homo* species more complex than others, the term complex is far too broad. To do so would be to take an overly homocentric approach to evolution and fail to appreciate complexities found within other species. Nonetheless, the unique intricate organization of our brain has provided us with cognitive abilities and problem-focused reasoning skills which surpass those of other species (Brown, 2001). Fundamentally, our sensory perception of the world and development of culture (i.e. language, art, music, etc.) must be attributed to our internal neurological mechanisms, which would not be possible without our higher cognitive capabilities (Brown, 2001). Furthermore, studies in the fields of psychology, sociology, and anthropology, among others, each indirectly involve an understanding of the brain. Therefore, for a comprehensive investigation into the psychological and cultural evolution of the *Homo* species, one must carefully consider the implications of varying brain morphology.

A Comparison Between *Homo Sapiens* and Early Primate Brain Morphology

Modern day humans have an average braincase capacity of 1,400 cm^3 (DeSilva & Lesnik, 2008). To understand the development of the current human brain, both comparative studies between humans and extant species as well as investigations of the fossil records have been undertaken. In general, previous research has identified the human brain to diverge from that of early primate ancestors on the basis of three distinguishing features: size, neocortex morphologies, and the quantity of cortex functional divisions (Kaas, 2012).

Our earliest bipedal human ancestors separated from the chimpanzee and bonobo lineage approximately 6 million years ago (mya) (Kaas, 2012). At this point in our history, our hominid ancestor's brain size was comparable to that of the great apes. Within the range of 400 to 600 cm^3, the hominid ancestors had a brain capacity approximately three times smaller than that of modern day humans (Kaas, 2012).

Nevertheless, within this time period, the brain had already begun to develop many of its intricacies that it retains today

(Kaas, 2012). For instance, hemispheric lateralization, while more defined in modern day humans, was prominent both within the *Homo* and the great ape lineages (Kaas, 2012). Hemispheric lateralization refers to unequal distribution of processes and functions attributed to each hemisphere of the brain (Silberman & Weingartner, 1986). Due to this, it is hypothesized that brain lateralization occurred longer than 6 mya and was naturally selected for ever since (Kaas, 2012). The expansion of brain size seen in the *Homo* species is primarily due to the addition to the neocortex, a region composing approximately 80% of *Homo sapiens'* total brain size (Kaas, 2012). However, as our brains evolved, each section of the neocortex did not develop evenly. In comparison to our early ape ancestors, the prefrontal, insular, posterior parietal, and temporal cortex all underwent expansion, developing more elaborate neuronal connections (Kaas, 2012). Additionally, Broca's area, which is specialized for language production, is also more sophisticated in current humans as compared to our early ancestors (Kaas, 2012). In contrast, the somatosensory and motor cortices of modern humans closely resemble those found in the early primate ancestors (Kaas, 2012). The somatosensory and motor cortices are brain locations attributed to touch and pain perception and voluntary movement respectively (Umeda et al., 2019). The development of a larger brain is metabolically costly and is associated with childbirth complications; therefore, the areas of the brain that saw considerable growth did not occur by random chance alone (Cofran & DeSilva, 2015). Instead, from the brain structures that underwent morphological changes in *Homo sapiens*, it can be deduced that higher level cognitive abilities were selected for (Cofran & DeSilva, 2015).

Fissures are collections of cerebral cortex tissue which compose indententations found on the surface of the brain (Kaas, 2012). The amount of brain tissue involved in each fissure varies among species, and provides us with another example in which human brains are different from those of other primates (Kaas, 2012). The depth and location of multiple fissures within the cortex vary among *Homo sapiens* and our hominin ancestors. However, the insula, located in a large fissure identified as the lateral sulcus, is of peculiar interest (Uddin et al., 2017). The insula of a *Homo sapiens* brain trumps that of primates both in size and in its relative proportion of the brain itself (Kaas, 2012). In both primates and humans, the insula is primarily concerned with perception of taste, pain, temperature, touch, and integration mechanisms (Uddin et al., 2017). However, in *Homo sapiens*, this region is also foundational to the limbic system providing us with emotional intelligence, social awareness and empathetic competencies (Kaas, 2012). Again, from these findings, one can deduce that emotional intelligence must have been vital to the *Homo* genus' survival over evolutionary history.

Neurological Evolution and Development

Australopithecus afarensis

Following the *Homo* species' divergence from chimpanzees and bonobos approximately 6 mya, human ancestors have undergone extensive changes in brain morphology (Kaas, 2012). The first bipedal human ancestor *Australopithecus afarensis*, commonly identified as Lucy, is thought to have walked on the planet approximately 3.4 million years ago in eastern Africa (Gibbons, 2020). With an average braincase capacity of 440 cm^3, Lucy's species' brain was identified to be 20% larger than those of chimpanzees (Gibbons, 2020). Yet, it lacked the neuronal density found in modern day humans (Gibbons, 2020). Such a finding indicates brain tissue growth precedes the development of neuronal circuitry (Gibbons, 2020). In turn, this questions the long-standing hypothesis that brain size is correlated to intellectual capabilities (Gibbons, 2020). It appears that no clear and direct correlation between the two can be identified.

Additional studies of 'Lucy's baby,' known as the Dikika child, have permitted paleontologists and anthropologists alike to begin to comprehend the developmental process of brain formation. In their research, paleontologists identified the recovered skull of the 2.4 year old Dikika child's brain and the brain of a chimpanzee to be of comparable size (Gibbons, 2020). Therefore, to reach the adult brain size of *A. afarensis*, the Dikika child must have been subjected to an overall longer duration of brain growth (Gibbons, 2020). As morphological brain growth occurs concurrently with cognitive development, scientists have since been fascinated with postulations on typical behavioral patterns that may have been correlated to *A. afarensis* (Neubauer & Hublin, 2012). However, these questions remain largely unanswered due to the limited quantity of recovered *A. afarensis* cranial specimens, which only included the remains of two juveniles and five adults (Neubauer & Hublin, 2012).

Homo erectus

Entering the *Homo erectus* stage of human evolution (2 million to 200,000 years ago), brain size underwent rapid changes. Over the course of only 800,000 years the cranial capacity nearly doubled (Hahn, 2012). As a result, discovered Javenese and Chinese *Homo erectus* specimens had an average braincase capacity of 930 and 1,029 cm³, respectively (Hahn, 2012). This measurement is only slightly smaller than the 1,350 cm³ braincase capacity of modern day *Homo sapiens* (DeSilva & Lesnik, 2008).

Unearthed in 1936, the Mojokerto calvaria, the only highly preserved skull of a *Homo erectus* child, began to provide scientists with some answers to the methodology of brain evolution (Hahn, 2012). This specimen remains a vital part of our current understanding of both brain development and evolution in the *Homo* phylogenetic lineage (Hahn, 2012). Nevertheless, the findings published on the implications of the size of Mojokerto calvaria are inconclusive (Hahn, 2012). While some analyses indicate that Mojokerto calvaria had an estimated death age of one year, others disagree (Hahn, 2012). In opposition, other published papers each suggest a slightly different estimated death age falling in the range of 1-6 years (Hahn, 2012). This broad potential death age range results in various proposed hypotheses of the cognitive abilities and fetal neurological development patterns of the *Homo erectus* species.

If the specimen had a death age of one year, some researchers conclude that members of the *Homo erectus* species experienced brain development much like chimpanzees, not yet developing secondary altriciality (Coqueugniot et al., 2004). Secondary altriciality refers to the human pattern of fetal brain development occurring post-parturition (Hahn, 2012). Unique to modern humans and some hominid ancestors, fetal brain development appears to rapidly continue within the first two years of life (Hahn, 2012). This period is thought to be critical for the development of cognition as white matter (myelinated neurons) expand rapidly, which increases brain mass (Hahn, 2012).

Still, other researchers contest this conclusion and instead support that the fetal brain development was inconsistent with patterns observed in *Homo sapiens* and in great apes (O'Connell & DeSilva, 2013). Instead, such researchers propose that the fetal brain development of *Homo erectus* should not be categorized with another species; rather, it should be seen as an independent mechanism of neurological development (O'Connell & DeSilva, 2013). Nevertheless, as *Homo erectus* did not have the fetal brain development pattern of *Homo sapiens*, it is hypothesized that *Homo erectus* may have lacked many of the higher cognitive abilities that characterize humans today (Coqueugniot et al., 2004). Due to the limited quantity of recovered *Homo erectus* specimens, specifically infant calvaria, conclusions of brain development are far from concrete. However, described fetal brain growth patterns have begun to shed light on the elusive evolutionary history of brain development.

Homo floresiensis

Since its introduction in 2004, the brain size of LB1, the type specimen of *Homo floresiensis*, has become a point of controversy within the scientific community. Moreover, this conflict has prevented a concrete classification of *Homo floresiensis'* status (Kubo et al., 2013). In the years following *Homo floresiensis'* initial discovery, several labs undertook the task of determining the endocranial volume (ECV) of LB1. Early published data, which utilized both traditional seed displacement methods and CT scans, indicated the ECV measurement to lie in the wide range of 380-430 cm^3 (Kubo et al., 2013). However, more recently, further studies with micro-CT scans have pointed to an ECV of LB1 of approx. 426 cm^3 (Kubo et al., 2013). Despite advancing technologies, exact measurements of LB1's ECV have yet to be pinpointed due to distortions, missing pieces, and the fusion of additional sediments to the specimen—all of which undermine the validity of ECV measurements (Kubo et al., 2013).

LB1's ECV measurements clearly diverge from the evolutionary trend of increasing brain size. With a braincase capacity of 380 to 430 cubic centimeters, *Homo floresiensis*' brain is estimated to be slightly smaller than that of *A. africans* (440 cm^3) (Kubo et al., 2013). While this finding was surprising by its own accord, associations of *Homo floresiensis* fossils with charred animal bones implied their ability to harness fire for cooking, further complicating the matter (Falk, 2005). Controversy remains as to whether the small brain size capacity would have allowed for the expression of such higher level cognitive abilities (Falk, 2005). Some scientists believe *Homo floresiensis* to be a species related to *Homo erectus*; however, others disagree with this classification (Falk, 2005). Instead, another long standing hypothesis is that *Homo floresiensis* was not in itself another species, but rather a human microcephalic (Falk, 2005).

In attempts to resolve the conflict of these opposing hypotheses, researchers created endocasts (reconstructions of internal braincases) to compare the brain of Homo *floresiensis* to other hominid ancestors and primates (DeSilva & Lesnik, 2008). All of the reconstructed endocasts were scaled to approximately 417 cubic centimeters to allow for a more fluid comparison of the brain structures (Falk, 2005). From these studies, it was established that LB1's brain structure most closely resembled that of *Homo erectus* from Java and China (Falk, 2005). Nonetheless, differences between the two are apparent. For instance, the occipital lobe of LB1 did not show expansion over the cerebellum as is prevalent in *Homo floresiensis* (Falk, 2005). More interestingly, of all the endocasts compared, *Homo floresiensis* did not show the morphologic characteristics of a microcephalic human: a pointed frontal lobe, compact occipital lobe, and flattening of the posterior brain (Falk, 2005). From these findings, the scientific proposal that *Homo floresiensis* was a microcephalic human was weakened.

Upon further inspection, smaller details of the various regions of LB1's brain have given scientists insight into the potential capabilities of the species. In general, the relative size of the frontal lobe between primates and *Homo sapiens* is not considerably different (Falk, 2005). Despite this, Brodmann's area 10, which is located in the prefrontal cortex of humans and apes, is significantly enlarged within the *Homo sapiens* species (Falk, 2005). Brodmann's area of *Homo erectus* and other primates do not appear to have been subjected to a similar enlargement (Falk, 2005). This, along with other research, has led scientists to postulate that this enlargement in modern humans may be correlated with higher order thought patterns (Falk, 2005). As with the *Homo erectus* endocast, LB1 did not exemplify this morphological characteristic. However, LB1 did not conform with *Homo erectus*' morphology in this region either (Falk, 2005). Instead, distinctive expansions identified in the prefrontal cortex region of LB1 suggest that *Homo floresiensis'* cognitions may have been more developed than *Homo erectus* and other early hominin ancestors (Falk, 2005). Furthermore, the large temporal lobe found in modern humans was also reflected in LB1's structure (Falk, 2005).

Overall, current data suggest that *Homo floresiensis* was not merely a microcephalic of the *Homo sapien* nor of *Homo erectus* (Falk, 2005). As LB1 appears to have derived characteristics of *Homo erectus*, one may hypothesize there to be a phylogenetic linkage between the two species (Falk, 2005). Still, it appears unlikely that *Homo floresiensis* is a miniaturized descendent of *Homo erectus* due to its brain-to-body ratio highly resembling *Australopithecus afarensis'* (Falk, 2005). To account for the limitations of the prior hypotheses, additional proposals of *Homo floresiensis'* origin have since been brought forward. Among these, scientists have discussed the potential of a not yet identified common ancestor between both *Homo floresiensis* and *Homo erectus* (Falk, 2005). Other hypotheses involve the possibility that *Homo floresiensis* was subjected to island dwarfism, resulting in the odd brain-to-body ratio observed (Falk, 2005).

However, limited discoveries of *Homo floresiensis* fossils have prevented such hypotheses from being confirmed or rejected. Further discoveries are imperative to the characterization of *Homo floresiensis'* fetal development, among other traits, that allow for its proper placement in our evolutionary history. Perhaps, as the fossil record grows, the scientific community will be able to come to a consensus on the proper phylogenetic linkages relating *Homo floresiensis* to the *Homo* ancestral tree.

Chapter 10: Bipedalism and Lower-Body Biomechanics of *H. floresiensis* (Ann Ping)

Ever since the discovery of LB1 in 2003, scientists have debated explanations regarding its morphology. There are two theories with the most support; first, that *H. floresiensis* was a descendant of an early australopith-like hominin, and second, that *H. floresiensis* was an insular descendant of *Homo erectus* through the process of island dwarfing (van Heteren, 2012). Insular (or island) dwarfing is when specific ecological conditions on small islands favour smaller individuals (Brown et al., 2004). Several factors are thought to contribute to insular dwarfism, including reduced resource availability, reduced predators, and limited calorie supply. This would place animals who have lower energy requirements at an advantage (Brown et al., 2004). Whereas more primitive traits in the morphology of *H. floresiensis* point to an ancestry deeper than *H. erectus*, paedomorphic (juvenile) features relative to *H. erectus* point to possible island dwarfing (van Heteren, 2012). The problem is that upon examination of the lower body morphology, support for both theories can be found through biomechanical and evolutionary lenses. This chapter will first address how scientists were able to conclude that *H. floresiensis* was bipedal. Then, in context of their bipedalism and locomotion, it will review the research on *H. floresiensis'* lower body morphology and subsequent claims regarding its ancestry.

Hypotheses Regarding the Evolution of Bipedalism

Humans and other hominids are largely characterized by their bipedalism, a form of locomotion where an organism walks with its two rear limbs. There are many hypotheses regarding the evolution of bipedalism, some examples being the Throwing Hypothesis, the Reaching for Food Hypothesis, the Thermoregulation Hypothesis, and the Wading Hypothesis (Niemitz, 2010). The Throwing Hypothesis predicts that bipedalism was a trailing adaptation that developed as early hominids started to use tools and weapons more frequently (Niemitz, 2010); orthograde (upright) locomotion was a lot more favourable towards the frequent throwing of objects (Niemitz, 2010). The Reaching for Food Hypothesis states that, in a savannah, our ancestors were forced to reach for high growing foods, which would have forced them to stand with upright posture (Niemitz, 2010). The Thermoregulation Hypothesis describes bipedalism as having been an adaptation to reduce the amount of heat absorbed from direct solar radiation (Niemitz, 2010). Finally, the Wading Hypothesis postulates that because high quality food was more abundant on the shores than in fragmenting forests at the time, our monkey or ape ancestors would have been forced to not only stand up, but also walk, in shallow waters to successfully harvest food (Niemitz, 2010). When considering the big picture, it's most likely that several of these hypotheses were at least partially correct and produced a combined effect that contributed to the evolution of bipedalism (Niemitz, 2010).

Morphological Indicators of Bipedalism

It is reasonable to believe that effective bipeds would have common morphological characteristics that allow the two hind legs to support the animal's entire body weight (Kramer, 2009). With this in mind, what characteristics are consistent with bipedalism and how were paleoanthropologists able to deduce that *H. floresiensis* was bipedal? The most reliable indicator of bipedalism in hominids is the shape of the pelvis, or more specifically, the ilium, which is the uppermost and largest part of the pelvis. A short, broad, and retroflexed (bent back) ilium has shown to be characteristic of bipedalism (Blaszczyk & Vaughan, 2007). In contrast, quadrupedal (four-legged) apes have tall, flat pelves (Gruss & Schmitt, 2015). The reason for this distinction is that shorter and wider ilia lower the body's center of mass and allow a necessary curvature in the lower back, called lumbar lordosis, that helps balance the upper body over the pelvis (Gruss & Schmitt, 2015). Additionally, the retroflexion of the pelvis allows the outer gluteal muscles to balance the trunk efficiently during bipedal locomotion (Gruss & Schmitt, 2015). Without this gluteal action, the upper body would shift from side-to-side with each step as weight shifts from one leg to another—this costs more energy (Gruss & Schmitt, 2015).

The Femur

Examining further down the skeleton, the femoral head is small and the femoral neck is long (Blaszczyk & Vaughan, 2007). The femoral head is the highest and globular-shaped part of the femur and the femoral neck supports the head. These features are regarded as primitive, meaning that they are similar to those of australopithecines or even chimpanzees (Blaszczyk & Vaughan, 2007). Biomechanically, such a femoral head-neck proportion is assumed to place greater strain on the hip joint. Moreover, LB1's wide hips are also seen as features that would increase hip joint friction forces (Blaszczyk & Vaughan, 2007). In this sense, one may wonder how *H. floresiensis* were efficient bipeds if there was constant strain on the hip joints. The answer lies in the fact that while in isolation these two features appear to be poor adaptations to bipedalism, they actually act in opposite directions to reduce reaction forces (Blaszczyk & Vaughan, 2007). LB1 also has a relatively high bicondylar angle of 14°, which is consistent with that found in *Australopithecus* (Brown et al., 2004). The bicondylar angle is the angle made between the femur and the body's midline. A higher bicondylar angle means that the femur is more "tilted", with the distal end angled towards the body's midline. However, other researchers argue that a bicondylar angle of 14° is still at or only slightly beyond the range reported for modern humans (Jungers et al., 2009b). Moreover, due to the short femurs of *H. floresiensis*, a greater bicondylar angle is necessary to place the feet closer to the body's center of mass, which allows for efficient bipedal gait (Jungers et al., 2009b). Thus, it is debatable whether the high bicondylar angle is an indicator of primitive morphology or if it is simply an adaptation to bipedal gait in a primate with very short legs. This, in combination with the fact that the distal end of the femurs in LB1 were extensively damaged (Jungers et al., 2009b), makes it difficult to determine whether femoral morphology supports the insular dwarfing theory or if it supports an ancestry deeper than *H. erectus*.

In fact, Vančata (2005) found that despite its shortness and robustness, the shape of the femurs are *Homo*-like, except more similar to Neanderthals than to *H. erectus*. They conclude that *H. floresiensis* may not be a simple example of island dwarfism, but a result of more complex phylogenetic processes. Overall, interpretations of LB1's femoral morphology seem to differ at least slightly across the literature. However, the nuances present two conclusions. Firstly, it may not be suitable to use femoral morphology in phylogenetic investigations because the femurs (and the pelves) play an important role in bipedalism. Secondly, the femoral morphology has a mosaic of primitive, modern, and unique features, making the phylogeny of *H. floresiensis* not something easy to determine.

The Tibia

Below the femurs are the tibiae. Like the femurs, the tibiae are also short, hence *H. floresiensis*'s nickname, 'hobbit human'. The ratio of the tibia to femur is *Homo*-like (Vančata, 2005), but the shape of the tibia is, interestingly enough, distinct from *Homo* (Brown et al., 2004). It is more robust (i.e. the diameters are large for the length), falling within the chimpanzee range of variation (Brown et al., 2004). In the context of bipedalism, robust and shortened limbs pull the body's center of gravity closer to the ground, which increases stability (van Heteren, 2012). Many island fauna that have experienced insular dwarfism also exhibit robust and shortened limbs (van Heteren, 2012). Once again, *H. floresiensis* exhibits a mixture of primitive and modern morphological features that play largely into their method of locomotion.

The Feet and Toes

The feet of *H. floresiensis* are very long relative to the femur and tibia (Jungers et al., 2009a). This proportion is primitive; the relative foot length is longer than that of humans and *Australopithecus afarensis*, and is consistent with those of bonobos (Jungers et al., 2009a). Not only are the feet relatively long, but the lateral toes (toes except the hallux, or the big toe) of *H. floresiensis* are also disproportionately long compared to modern humans, making the forefoot very long (Jungers et al., 2009a). Specifically, the proximal pedal phalanges, the most proximal set of bones out of the three sets of bones that make up the toes, are longer than and lack the hourglass shape found in those of modern humans (Jungers et al., 2009a). Furthermore, the length of the proximal pedal phalanges resembles those in some chimpanzees, and its toes are more curved than those of modern humans, resembling those of some australopithecines (Jungers et al., 2009a). Among the tarsal bones (the seven bones located proximally in the foot, near the ankle), the navicular is the most primitive, exhibiting a wedge shape that is seen in australopithecines and great apes (Jungers et al., 2009a). The talus, another tarsal bone, is regarded as 'intermediate' in shape, in that it has some human-like features and some ape-like features (Jungers et al., 2009a). Interestingly, the proportions of the feet and the shape of their bones are overwhelmingly primitive in morphology. This differs from what has been observed with the femurs and tibiae. Hence, *H. floresiensis*'s unique foot morphology seems to support the theory that the species was a descendent of a primitive hominin.

Locomotion in *H. floresiensis*

Previously, the lower body morphology of *H. floresiensis* was discussed in relation to their phylogeny. Now, the established information will be consolidated using a biomechanical and kinematic lens to review how the lower body morphology of *H. floresiensis* influenced walking and running. When considering the energy requirements of short legs, it may seem like short legs increase energy expenditure because of their short stride length, which would result in a larger cadence (steps per minute) for a given velocity (Blaszczyk & Vaughan, 2007). A large cadence means that the legs swing more often for a given velocity, resulting in greater energy expenditure. However, Kramer and Eck (2000) claim that such reasoning is not completely correct, as it fails to recognize that longer legs also have a higher mass moment of inertia, or in other words, longer legs have a higher resistance to acceleration. This is because more mass, located in the foot, is distributed further from the center of rotation at the pelvis. This would increase the power requirements of locomotion.

In fact, Steudel (1994) finds that there is no significant relationship between the energetic cost of locomotion and limb length in 21 species of mammals. If the discussion so far were to be extended to *H. floresiensis*, it would be difficult to determine how their lower body morphology impacts their metabolic costs of locomotion with leg length being the sole factor of interest. Fortunately, there is substantial research on the details of the foot morphology of *H. floresiensis* which can provide more insight to this discussion. The feet of *H. floresiensis* are unique and fundamental to an understanding of their locomotion. It has been found that increasing distal limb mass produces substantial metabolic spendings during endurance running, but not during walking, for which there is little effect (Bramble & Lieberman, 2004). It is reasonable that increasing distal limb mass would be more costly because it also increases the mass moment of inertia. Hence, from a 'toe length' point of view, increasing toe length would increase distal limb mass, increasing
the cost of locomotion.

Additionally, Rolian et al. (2009) predicted that the digital flexors exert greater forces and do more mechanical work during running to prevent the metatarsophalangeal (MTP) joints from collapsing into hyperextension during propulsion. The digital flexors are muscles in the lower leg and feet that are involved in flexion of the feet and toes and the MTP joints are the joints located at the proximal end of the toes. There is a greater load placed on the forefoot during running compared to walking, and these loads are placed on the distal ends of the toes. To stabilize the MTP joints which are bearing these loads, the digital flexors must exert greater forces and do more work (Rolian et al., 2009). Rolian et al.'s (2009) model was ultimately supported by their results, which showed that there was indeed an increased mechanical cost associated with long toes in running.

The foot of *H. floresiensis* also lacks a well-defined medial longitudinal arch (the arch of the foot when viewed from the inner side), which limits the recovery of stored elastic energy in the foot. This is especially relevant during running, when mass-spring mechanics come into play (Jungers et al., 2009a). Finally, because the feet of *H. floresiensis* were so long, there would also have been kinematic adjustments made to the ankle, knee and/or hip joint to allow proper clearance of the foot, which would have been evident during walking and running (Jungers et al., 2009a). Overall, the biomechanical effects of short legs, long feet, long toes, and a poorly-defined medial longitudinal arch suggest that the lower body morphology of *H. floresiensis* may not have been designed well for walking, and even more poorly designed for running.

Conclusion

This chapter addressed several important points regarding the lower body morphology of *Homo floresiensis* in relation to their bipedalism, phylogeny, and locomotion. Despite slight differences in interpretation of the phylogeny of *H. floresiensis* in the literature, the common conclusion is that the lower body morphology of H. floresiensis presents a mosaic of primitive, modern, and unique features. Additionally, despite seemingly primitive features, we are cautioned that they could potentially be adaptations to bipedalism, and that assuming that they were features derived from *Australopithecus* or other genera primitive to *Homo* may be fallacious. Finally, in an in-depth review of how all the lower body elements come together to influence locomotion, it was revealed that *H. floresiensis* were likely not the most optimal walkers or runners. The discovery of *Homo floresiensis* was an event that excited and confused paleoanthropologists all around the world. Amongst all the debates, one thing runs true: the fascination brought through 'simple' fossil remains of a creature that once lived 50 000 years ago is not something to be underestimated.

Chapter 11: Myths and Mysteries surrounding the *H. floresiensis*: What does this tell us about human fear of the 'other'? (Vedanshi Vala)

An introduction: Why are we afraid?

We are afraid. As a society, as a humanity, we fear the outliers. The unknown terrifies us, keeps us awake at night as we watch the symphony of our fears paralyze us in our nightmares. We are the same humans whose curiosity has taken us to many places in the universe, from walking steps on the moon, to probing our genetic makeup at the molecular level with advancements in CRISPR technology (Broad Institute, 2021). If we are to mull over our tendency to ostracize everything we do not understand, our habit of prejudicing before experiencing, perhaps the origins of horror stories can be understood to be ourselves. We are the masters wielding the very strings that enslave our puppet minds to our own version of the truth. What is true in this world? A scientist may claim the truth is that which can be studied through an empirical, systematic process. A person with a different worldview may claim that the truth lies in supernatural phenomena, which cannot be studied through the scientific method. This frames the approach of this exploration, which seeks to investigate the human fear of the unknown and its connection to the origins of mythology, and the differentiation between empirical truth and that presented by the supernatural in an anthropological context. Of particular interest will be folklore surrounding *H. floresiensis*, with commentary on the validity and potential for co-existence of various worldviews.

What is true?

A brief comparison between the scientific and supernatural worldviews

To contextualize this discussion, there must first be further elaboration upon the two worldviews of interest: the scientific worldview, otherwise known as the empirical worldview, and the supernatural worldview. The scientific investigative process supports a naturalistic worldview, where any claims must be evaluated through a framework which can produce testable scientific evidence (Fishman, 2009). Conversely, the supernatural worldview is founded on other elements, such as paranormal or enchanted forces or greater powers, or in other words, that which cannot yet be understood or investigated by the current scientific paradigm (Fishman, 2009). Between two seemingly opposite worldviews, where does the truth about this world lie?

Consider that the images seen, sounds heard, or sensory experiences otherwise perceived by a person can be stimulated through a series of inputted electrical impulses (Penfield, 1958). The 'brain in a vat' hypothetical model argues the possibility of the brain being inside a container, rather than inside an individual, and being supplied with information to fabricate the semblance of living without the presence of its body (Gere, 2004). If this were the case, there would be no definitive manner with which the truth of an individual's existence can be disputed using empirical logic or deductive reasoning. However, supernatural worldviews have other ways of explaining human existence, such as through mythology in various cultures providing in-depth accounts of the origins of the human race; specifically, etiology refers to myths which explain origins (Birrell, 1999). There is evidently an opposing nature between how a scientist would approach this puzzle versus an individual with traditional ways of understanding the world, and as such, one would find themselves at an impasse.

The question that remains is whether these two worldviews, the empirical and the supernatural, can be consolidated. It can, however, be argued that a more collaborative approach needs to be taken to approach this debate. In order to determine the truth, then, one could argue that both the supernatural and empirical worldviews can be used to understand the reality in which the inquirer finds themselves. This manner with which to restructure one's understanding of that which is real in the world, and the worldviews with which the truth can be approached, frame the necessary mindset to explore the topics discussed in the remainder of this chapter.

The legend of the ebu gogo

And how myths originate

Prior to discussing the origins of mythology, like that about the ebu gogo, it is valuable to understand the scientific background behind the species classification of *H. floresiensis*, on whose existence such legends are based. *Homo floresiensis* is considered to be a human species that is differentiated from the taxonomic classification of *Homo sapiens*, and physically are distinguished from *H. sapiens* through their significantly different body proportions, such as shorter height and larger teeth (Smithsonian Institution, 2020). It has been theorized that these remarkable differences are the result of a phenomenon known as island dwarfism, which is a process of evolution through which genetic differences are developed in light of food scarcity and lower risk of predation (Smithsonian Institution, 2020). Given this background, one can begin to unravel the many complexities embedded within the myth of the ebu gogo.

The legend of the ebu gogo, originating from Flores in Indonesia, is as follows: the ebu gogo is the forest's grandmother, having an appetite for everything ranging from crops to human flesh (Madison, 2020). It is further documented in folklore that at a certain point in time, there roamed beings in those very forests who were small in size, with hairy bodies, who lived alongside local *H. sapiens* and consumed a diet akin to that of the ebu gogo (Madison, 2020). Until the relatively recent discovery of bones belonging to the species now classified as *H. floresiensis*, lore surrounding the ebu gogo and other such tiny people were considered to be the product of supernatural worldviews; however, the new discovery broke open a chasm of possibilities regarding the value of such mythology to scientific investigations (Madison, 2020).

Moreover, there is a conversation to be had regarding the origins of mythology and folklore, and their reflection on the boundaries that confine reality within certain worldviews. Returning to the roots of certain myths, some are motivated by a need to understand and explain the natural world, its workings, and its origins (Birrell, 1999). For instance, Greek mythology, beyond documenting exciting supernatural phenomena, also served to elucidate "such monumental events as the creation of human beings" (Payment, 2006, p. 9). This exemplifies how folklore is used to explain happenings through a supernatural worldview.

As such, the discovery of *H. floresiensis* and the evident overlap with traditional folklore brings about fascinating discussions on whether such tales are indeed grounded in empirical truth. However, in encountering this perspective, the academic community needs to consider whether an idea needs to have an empirical foundation in order to be true. The example of the legend of the ebu gogo demonstrates how traditional ways of explaining the world far preceded scientific discovery about an entirely different taxonomic group's co-existence with *H. sapiens*. In sum, simply because a supernatural phenomenon is not supported through empirical evidence, its validity doesn't need to be entirely discredited. Similarly, if science is able to demonstrate significant advancements in a particular sector which is as of yet not explained by traditional worldviews, it should not call for lack of trust in the result of the empirical process. Ultimately, mythology is simply one manner with which the world can be understood — it cannot be taken as the whole truth, nor can it be scoffed upon simply because of its differences from the scientific worldview.

Understanding human fear

A conversation on diversity, prejudice, and racism

Myths and folklore, as discussed in the previous section, are the products of an innate human desire to explain mysterious phenomena. By dissecting this further, it can be hypothesized that such phenomena induce significant intrigue and fear, thereby resulting in an ensuing need to protect oneself or others from a threat whose nature is not yet fully understood. In other words, when a person or group's existing realm of knowledge fails to explain their experience, they would attempt to develop a rationale lying outside the boundaries of empirical logic. With this in mind, a point of discussion that warrants attention is whether assumptions of certain phenomena being other-worldly or supernatural in nature are the products of inherent bias, prejudice, and racism.

That claim is rather bold; however, there is certain evidence yielding itself to such an assertion. Take the genre of science fiction as a whole, which, while not explicitly falling under the category of mythology, can be studied through a similar lens to understand human fear of the 'other'. Science fiction entails popular media such as movies, television shows, and comic books, set in a hypothetical world at a future time. Given its nature, this media can divulge significant insight into the creators' views regarding the place held by certain racial groups in society in the current real world, and where they are predicted to be in the future should events follow a certain pattern or trajectory (Por, 2005). Moreover, such media plays a role in defining who the 'other' is relative to the protagonists of the plot, where the 'other' are often people of colour who are cast into alienating roles (Por, 2005). This includes choices on who is portrayed as a villain versus the hero, who is shown to be weak, and who has some kind of supernatural abilities enabling far greater strength in that universe (Por, 2005). "The vast majority of films in the genre are about fighting the evil aliens who threaten our very existence", with the heroic roles being white dominated, and people of colour being cast into villainous roles (Por, 2005, p. 5). Moreover, it is often the nature of science fiction to attribute magnificent creations from ancient civilizations to aliens—consider that this sort of mythology, in the form of conspiracy theories, is perpetuated because of a white supremacist or eurocentric worldview which refuses to credit other, 'primitive', groups for advancements in science and technology (Reynolds, 2015). These science fiction stories, myths, and conspiracy theories are revealing of a mindset that considers non-white racial groups to be inferior or backwards. This demonstrates how a lack of understanding of the 'other', the propensity to see the world through a narrow field of vision, and being blinded by prejudice is the catalyst driving such harmful narratives.

These examples demonstrate how stereotyping and discriminatory practices are perpetuated by myths, heightening the marginalization of already vulnerable ethnic groups. In the face of such a reality, one may come to ask oneself whether diversity as a whole should be ignored, in the sense that despite differences and varying needs, everyone is treated equally. Another alternative, and perhaps one which is a better approach to this blanket ignorance of diversity, would be to increase intercultural understanding and respect, thereby working towards equity rather than equality. Through such a system, differences could be celebrated rather than ignored, and the needs of different individuals and groups would be assessed independently of what another group or person's requirements are. Diversity is seen through the many labels society enforces on people, be it in body size, shape, height, or the colour of skin, hair, eyes, or nuances of speech and language, cultural attire, traditions, food, and lifestyles, or a person's sexuality, gender identity, and other such aspects that contribute to an individual's external identity in the eyes of society. When dismembering identity into a list of labels, as with the previous sentence, it can alienate humans from each other; however, taken holistically, they are simple nuances of what humanity is. People are truly, at a fundamental level, the same— even if *H. sapiens* and *H. floresiensis* are taxonomically different, it is on a shared Earth that both groups coexist. Compassion and a willingness to learn and celebrate such differences, and the courage to remain unified despite them, is unifying. The human experience is unifying, and can conquer fears, break barriers, and solve mysteries that would otherwise alienate people from each other.

Conquering our minds - a conclusion

Finding the truth about the human experience

How does one even begin to discuss fear? While there is no single approach to the exploration of such a topic, this chapter was undertaken through the lens of the mythology surrounding *H. floresiensis*.

As a refresher, this chapter discussed firstly the differentiation between the empirical and supernatural worldviews, and deliberated upon whether they may be reconciled in an effort to understand the true reality of human existence. Then, the afore-mentioned arguments were used to delve into the origins of mythology within the context of the tale of the ebu gogo. This exploration was concluded through a deep-dive into a theme as bold and palpable as human fear, realizing in the process how human curiosity must be treated delicately or risk societal calamities. The objective of this exploration was to learn about the truth behind the human experience of fear from the example of *H. floresiensis*. As it was concluded, such fear is born out of trepidation surrounding that which is unknown, or in other words, that what makes people uniquely different from each other. Cherishing such diversity requires intercultural understanding and respect, but more so an unbiased heart, an open mindset, and a willingness to alter course after being wrong. In all, despite the very many differences between one person to the next, it can be realized that humankind's intriguing, fascinating, and impossible existence warrants celebration; the human race's many mysteries, fears, and truths are all part of what comprise the human experience.

Yes, the human race is afraid of the unknown. However, the moment it is endeavored to introduce compassion in the midst of uncertainty, the moment hands are extended in friendship rather than to cup over a friend's ear and whisper riveting tales, the moment humans see each other for who they are rather than what prejudice dictates, people can conquer their self-created fear. Each person has the power to conquer their own mind. As people understand the myth of the ebu gogo, and countless others like it, to be a construction of human fear of the unknown, they can liberate themselves from the very fears they have imprisoned their minds in.

References

Bjørlykke, K. (2014). Relationships between depositional environments, burial history and rock properties. Some principal aspects of diagenetic process in sedimentary basins. *Sedimentary Geology*, *301*, 1–14. https://doi.org/10.1016/J.SEDGEO.2013.12.002

Gutiérrez, Y., Ott, D., Töpperwien, M., Salditt, T., & Scherber, C. (2018). X-ray computed tomography and its potential in ecological research: A review of studies and optimization of specimen preparation. *Ecology and Evolution*, *8*(15), 7717. https://doi.org/10.1002/ECE3.4149

Immel, A., Cabec, A. le, Bonazzi, M., Herbig, A., Temming, H., Schuenemann, V. J., Bos, K. I., Langbein, F., Harvati, K., Bridault, A., Pion, G., Julien, M.-A., Krotova, O., Conard, N. J., Münzel, S. C., Drucker, D. G., Viola, B., Hublin, J.-J., Tafforeau, P., & Krause, J. (2016). Effect of X-ray irradiation on ancient DNA in sub-fossil bones – Guidelines for safe X-ray imaging. *Scientific Reports 2016 6:1*, *6*(1), 1–14. https://doi.org/10.1038/srep32969

Lautenschlager, S. (2016). Reconstructing the past: methods and techniques for the digital restoration of fossils. *Royal Society Open Science*, *3*(10). https://doi.org/10.1098/RSOS.160342

Šejnoha, J., Šejnoha, J., Jarušková, D., Špačková, O., & Novotná, E. (2021). *Risk Quantification for Tunnel Excavation Process*. Unpublished manuscript from http://citeseerx.ist.psu.edu/viewdoc/summary?doi=10.1.1.308.8848

Waagen, J. (2019). New technology and archaeological practice. Improving the primary archaeological recording process in excavation by means of UAS photogrammetry. *Journal of Archaeological Science, 101*, 11–20. https://doi.org/10.1016/J.JAS.2018.10.011

Wagner, J. K., Yu, J.-H., Ifekwunigwe, J. O., Harrell, T. M., Bamshad, M. J., & Royal, C. D. (2017). Anthropologists' views on race, ancestry, and genetics. *American Journal of Physical Anthropology, 162*(2), 318–327. https://doi.org/10.1002/AJPA.23120

Wallerstein, by I. (2015). Anthropology, Sociology, and Other Dubious Disciplines. *The University of Chicago Press Journals, 44*(4), 453–465. https://doi.org/10.1086/375868

Argue, D., Groves, C., Lee, M., & Jungers, W. (2017). The affinities of Homo floresiensis based

on phylogenetic analyses of cranial, dental, and post-cranial characters. *Journal of Human*

Evolution, 107, 107–133. https://doi.org/10.1016/J.JHEVOL.2017.02.006

Baab, K. (2016). The place of Homo floresiensis in human evolution. *Journal of Anthropological*

Sciences = Rivista Di Antropologia : JASS, 94, 5–18. https://doi.org/10.4436/JASS.94024

Baab, K. L. (2012). Homo floresiensis: Making Sense of the Small-Bodied Hominin Fossils from

Flores. *Nature Education Knowledge, 3*(4).

Gómez-Robles, A. (2016). The dawn of Homo floresiensis. *Nature 2016 534:7606, 534*(7606),

188–189. https://doi.org/10.1038/534188a

de Tienda Palop, L., & Currás, B. X. (2019). The Dignity of the Dead: Ethical Reflections on the Archaeology of Human Remains. In *Ethical Approaches to Human Remains*. https://doi.

org/10.1007/978-3-030-32926-6_2

Galway-Witham, J., & Stringer, C. (2018). How did Homo sapiens evolve? In *Science* (Vol. 360, Issue 6395). https://doi.org/10.1126/science.aat6659

González-Ruibal, A. (2018). Ethics of Archaeology. *Annual Review of Anthropology*, *47*(1). https://doi.org/10.1146/annurev-anthro-102317-045825

Joannes-Boyau, R., Pelizzon, A., Page, J., Rice, N., & Scheffers, A. (2020). Owning humankind: fossils, humans and archaeological remains. *Heliyon*, *6*(6). https://doi.org/10.1016/j.heliyon.2020.e04129

Overholtzer, L., & Argueta, J. R. (2018). Letting skeletons out of the closet: the ethics of displaying ancient Mexican human remains. *International Journal of Heritage Studies*, *24*(5). https://doi.org/10.1080/13527258.2017.1390486

Sallam, A. (2019). The ethics of using human remains in medical exhibitions: A case study of the cushing center. *Yale Journal of Biology and Medicine*, *92*(4).

Scarre, G. (2003). Archaeology and respect for the dead. *Journal of Applied Philosophy*, *20*(3). https://doi.org/10.1046/j.0264-3758.2003.00250.x

Scarre, G. (2014). 7. The Ethics of Digging. In *Cultural Heritage Ethics: Between Theory and Practice*. https://doi.org/10.11647/obp.0047.07

Swain, H. (2002). The ethics of displaying human remains from British archaeological sites. *Public Archaeology*, *2*(2). https://doi.org/10.1179/pua.2002.2.2.95

The Norwegian National Research Ethics Committees. (2016). Guidelines for Research Ethics, in the Social Sciences, Humanities, Law and Theology. In *SSRN Electronic Journal*.

Cavalli-Sforza, L. L., & Feldman, M. W. (2003). The application of molecular genetic approaches to the study of human evolution. *Nature Genetics*, *33*(S3), 266–275.

https://doi.org/10.1038/ng1113

Gasperskaja, E., & Kučinskas, V. (2017). The most common technologies and tools for

functional genome analysis. *Acta Medica Lituanica*, *24*(1), 1–11.

https://doi.org/10.6001/actamedica.v24i1.3457

Hedrick, P. W. (2011). Population genetics of malaria resistance in humans. *Heredity*, *107*(4),

283–304. https://doi.org/10.1038/hdy.2011.16

Kappelman, J. (1996). The evolution of body mass and relative brain size in fossil hominids.

Journal of Human Evolution, *30*(3), 243–276. https://doi.org/10.1006/jhev.1996.0021

Novembre, J., & Di Rienzo, A. (2009). Spatial patterns of variation due to natural selection in

humans. *Nature Reviews Genetics*, *10*(11), 745–755. https://doi.org/10.1038/nrg2632

Falk, D. (2005). The Brain of LB1, Homo floresiensis. *Science*, *308*(5719), 242–245.

https://doi.org/10.1126/science.1109727

U.S. Department of Health and Human Services. (n.d.). *Laron syndrome*. Genetic and Rare

Diseases Information Center.

https://rarediseases.info.nih.gov/diseases/6859/laron-syndrome.

Kubo, D., Kono, R. T., & Kaifu, Y. (2013). Brain size of Homo floresiensis and its evolutionary

implications. *Proceedings of the Royal Society B: Biological Sciences*, *280*(1760),

20130338. https://doi.org/10.1098/rspb.2013.0338

Than, K. (2021, May 3). *Hobbit's Brain Size Holds Clues About Its Ancestor*.

https://www.nationalgeographic.com/pages/article/130418-hobbit-homo-floresiensis-brai

n-size-hominin-human-evolution#:~:text=Using%20a%20new%20high%2Dresolution,to

%20be%20about%20426%20cc.

Hershkovitz, I., Kornreich, L., & Laron, Z. (2007). Comparative skeletal features betweenHomo

floresiensis and patients with primary growth hormone insensitivity (Laron syndrome).

American Journal of Physical Anthropology, *134*(2), 198–208.

https://doi.org/10.1002/ajpa.20655

Baab, K. L., McNulty, K. P., & Harvati, K. (2013). Homo flore-siensis Contextualized: A

Geometric Morphometric Comparative Analysis of Fossil and Pathological Human

Samples. *PLoS ONE*, *8*(7). https://doi.org/10.1371/journal.pone.0069119

Falk, D., Hildebolt, C., Smith, K., Morwood, M. J., Sutikna, T., Jatmiko, Saptomo, E. W., Imhof,

H., Seidler, H., & Prior, F. (2007). Brain shape in human microce-phalics and *Homo*

floresiensis. *Proceedings of the National Academy of Sciences*, *104*(7), 2513–2518.

https://doi.org/10.1073/pnas.0609185104

Vannucci, R. C., Barron, T. F., & Holloway, R. L. (2011). Cranio-metric ratios of microcephaly

and LB1, *Homo floresiensis*, using MRI and endocasts. *Proceed-ings of the National*

Academy of Sciences, *108*(34), 14043–14048. https://doi.org/10.1073/pnas.1105585108

Chahine, G., Diekhof, E. K., Tinnermann, A., & Gruber, O. (2015). On the role of the anterior prefrontal cortex in cognitive 'branching': An fMRI study. *Neuro-psychologia, 77,*

421–429. https://doi.org/10.1016/j.neuropsychologia.2015.08.018

Walters, C. (2020). The geometry of abstraction in hippocampus and pre-frontal cortex.

https://doi.org/10.1242/prelights.17378

Martin, K. A. (2004). Faculty Opinions recommendation of A new small-bodied hominin from

the Late Pleistocene of Flores, Indonesia. *Faculty Opinions – Post-Publication Peer*

Review of the Biomedical Literature. https://doi.org/10.3410/f.1020854.250610

www.ingramcontent.com/pod-product-compliance
Lightning Source LLC
Chambersburg PA
CBHW030852270326
41928CB00008B/1341